Cyanobacteria - Recent Advances in Taxonomy and Applications

Edited by Wael N. Hozzein

Published in London, United Kingdom

IntechOpen

Supporting open minds since 2005

Cyanobacteria - Recent Advances in Taxonomy and Applications
http://dx.doi.org/10.5772/intechopen.92398
Edited by Wael N. Hozzein

Contributors
Suman Kundu, Sheetal Uppal, Mohd Asim Khan, Gabriel Komla Ameka, Ayesha Algade Amadu, Kweku Amoako Atta deGraft-Johnson, Jian Yuan, Kyoung-Jin Yoon, Laxmi Parwani, Mansi Srivastava, Jaspreet Singh, Taiara A. Caires, Helen Michelle de J. Affe

First published in London, United Kingdom, 2022 by IntechOpen
IntechOpen is the global imprint of INTECHOPEN LIMITED, registered in England and Wales, registration number: 11086078, 5 Princes Gate Court, London, SW7 2QJ, United Kingdom
Printed in Croatia

British Library Cataloguing-in-Publication Data
A catalogue record for this book is available from the British Library

Additional hard and PDF copies can be obtained from orders@intechopen.com

Cyanobacteria - Recent Advances in Taxonomy and Applications
Edited by Wael N. Hozzein
p. cm.
Print ISBN 978-1-83962-489-6
Online ISBN 978-1-83962-490-2
eBook (PDF) ISBN 978-1-83962-491-9

For EU product safety concerns:
IN TECH d.o.o., Prolaz Marije Krucifikse Kozulić 3, 51000 Rijeka, Croatia, info@intechopen.com or visit our website at intechopen.com.

Meet the editor

Prof. Wael N. Hozzein is a Professor of Microbiology at the Faculty of Science, Beni-Suef University, Egypt. He received his Ph.D. from Cairo University, Egypt, and went on to work as a visiting scientist at Newcastle University, UK, and Michigan State University, USA. He is the chair professor of the Bioproducts Research Chair at King Saud University, Saudi Arabia. He has vast experience in bacterial taxonomy, microbial biodiversity, and biotechnological applications of bacteria. Prof. Hozzein is the author of more than 180 publications and a guest editor, editorial board member, and reviewer for several international journals. Recently, he was included in Stanford University's list of the top 2% most-cited scientists. He has been the principal investigator for several funded grants and has also received several awards, including the State Encouragement Prize in Biological Sciences in 2015. Prof. Hozzein has been involved in many academic activities and educational reform projects and initiatives. Recently, he served as an advisor to Nahda University President for Development, Research, and Quality.

Contents

Preface

Cyanobacteria are an interesting group of bacteria for their unique characteristics and potential biotechnological applications. They are very important for life on Earth because they are oxygenic organisms that also play crucial roles in the cycles of carbon, nitrogen, and oxygen. Interestingly, they have unique specialized cells called heterocysts for nitrogen fixation. Therefore, they are ideal model organisms for studying photosynthesis, nitrogen fixation, and other biological processes. In addition, cyanobacteria are well recognized for their potential for a variety of biotechnological applications. They were reported to produce a wide array of biologically active compounds that are attracting interest from the pharmaceutical industry for drug development. Also, they have various applications in agriculture as potential biofertilizers or in the industry for biofuel production, in addition to many other biotechnological applications.

Due to the increasing interest in this group of microorganisms, this book discusses recent advances in taxonomy and applications of cyanobacteria. It is a useful resource for students starting their research work on cyanobacteria, researchers interested in the recent advancements in their research field, and teachers involved in teaching topics related to cyanobacteria. The book contains five chapters.

Chapter 1 discusses the molecular methods applied for identifying freshwater toxigenic cyanobacteria, reviewing recent methods to rapidly and accurately identify toxic cyanobacteria, which is difficult due to the lack of discernable morphological difference between toxic and non-toxic strains within the same cyanobacterial species or genus. This information is important for students and researchers to understand the methods they are going to use in their research.

Chapter 2 outlines the diverse industrial applications of cyanobacteria in the pharmaceutical, agricultural, and health sectors. The applications include the production of bioplastics, biofuels, biofertilizers, foods, nutraceuticals, and pharmaceuticals. Additionally, the metabolic pathways that lead to the production of some important cyanobacterial bioactive compounds are outlined in this chapter along with examples of the commercial products from cyanobacteria currently available on the market.

Chapter 3 reviews the potential of cyanobacteria in wound healing. It is clearly shown that these organisms have immense potential to be utilized for the development of bioactive wound dressings.

Chapter 4 introduces a very interesting and special application for *Synechocystis* sp. PCC 6803, which is the production of a novel hemoglobin. The chapter sheds light on the structure–function relationship and potential applications of this novel hemoglobin. It has been reported that cyanobacterial hemoglobins have displayed unprecedented stability, unique heme coordination, and other properties that are not often observed in the globin superfamily. Therefore, this chapter provides an overview of the unique globin from *Synechocystis* sp. PCC 6803 and its biotechnological implications, including potential in the field of artificial oxygen carriers.

Chapter 5 highlights the diversity, distribution, and applications of cyanobacteria in the Brazilian coastline in order to fill a gap in cyanoflora knowledge. Recognition of the Brazilian cyanoflora contributes to the understanding of the functioning and monitoring of marine ecosystems and provides data for the construction of future public policies.

I would like to thank all the contributors for sharing their excellent research work and for their academic integrity. I would also like to thank the staff at IntechOpen, particularly Author Service Manager Ms. Sara Debeuc for her commitment, patience, and keen assistance in making this book possible.

Finally, I hope that readers will find this book interesting and use the information contained herein to positively contribute to the research community.

Wael N. Hozzein
Botany and Microbiology Department,
Faculty of Science,
Beni-Suef University,
Beni-Suef, Egypt

Chapter 1

Overview of PCR Methods Applied for the Identification of Freshwater Toxigenic Cyanobacteria

Jian Yuan and Kyoung-Jin Yoon

Abstract

Although cyanobacteria are essential microorganisms on earth, some cyanobacteria produce toxins known as cyanotoxins, threatening humans and animals' health. Hence, it is imperative to rapidly and accurately identify those toxic cyanobacteria. Unfortunately, traditional microscopic methods have limitations for accurate identification due to the lack of discernable morphological difference between toxic and non-toxic strains within the same cyanobacterial species or genus. In contrast, their genetic profiles are inherently conserved; therefore, nucleic acid-based assays can be more reliable for precise identification. Furthermore, molecular assays can provide high throughput and significantly reduce the turn-around time of test results. Such advantages make those assays a preferred method for rapid detection and early warning of potential toxicity. Toxigenic cyanobacterial species have synthetase genes (DNAs) for toxin production, which can be excellent marker genes. Numerous molecular assays targeting cyanotoxin synthetase genes have been developed for the identification of toxigenic cyanobacteria at various taxonomic levels. Polymerase chain reaction (PCR)-based assays are the most prevailing. Among different versions of PCR assays, the real-time quantitative PCR can be utilized to quantify the genes of interest in samples, fulfilling the purpose of both taxonomic recognition and biomass estimation. Reverse transcription (RT)-PCR assays can be used to detect transcripts (i.e., mRNAs) from toxin synthetase genes, probably enhancing the predictive value of PCR detection for toxin production from observed cyanobacterial species. Nevertheless, the utility of toxin synthetase gene- or its transcript-based PCR assays for routine cyanotoxin monitoring needs to be further evaluated on a large scale.

Keywords: cyanobacteria, cyanotoxins, toxin synthetase genes, molecular techniques, polymerase chain reaction

1. Introduction

Cyanobacteria are essential microorganisms on earth as they produce oxygen and account for a large part of primary aquatic productivity. Simultaneously, some freshwater cyanobacteria can produce various toxins, named cyanotoxins, some of which are potently poisonous to humans and animals. A well-known cyanotoxicosis in humans was reported from Brazil in association with medical malpractice in 1996. In this incident, 126 patients in a hemodialysis unit were affected, and 60 of them died due to using microcystin-contaminated water from a local reservoir. A cyanobacterial bloom was found in that reservoir concurrently [1]. Besides, there have been reports concerning human cyanotoxin poisoning by drinking water or via injury after contacting recreational water [2]. Apart from humans, numerous animal poisoning cases have also taken place because they can reach the unprocessed natural water directly so that the risk of being poisoned becomes higher. These cases involve livestock, pets, and wildlife [3–10].

Cyanobacterial blooms occurred more frequently in recent years, which may have been attributed to the aggravating eutrophication in freshwater and global warming. As such, cyanotoxin poisoning incidents have also been increasingly reported. Nowadays, freshwater cyanobacterial blooms have broader geographical and temporal impacts on local water bodies that act as vital municipal or agricultural water supplies. With the possibility of cyanotoxin contamination, humans and animals residing in surrounding areas continue to be threatened. Therefore, testing for toxic cyanobacteria or cyanotoxins is imperative for detection and preventive measures.

Although cyanobacteria can be observed under a microscope, their toxigenicity cannot be determined by microscopy because the toxigenic cyanobacteria do not have unique morphological characteristics. Some laboratories have adopted a testing strategy that combines microscopic observation and cyanotoxin detection to indicate the existence of toxigenic cyanobacteria in samples. Although this strategy may seem reasonable and pragmatic, it needs collaboration between chemical analysts and microalgal biologists to reach an agreement on the conclusion. Furthermore, it neglects the complex phenomena of the same toxin production by different species or genera, leading to an incorrect judgment of the truly culpable toxin producers.

Cyanotoxin testing has been in place. Yet, available tests have shortcomings. For example, commercial enzyme-linked immunosorbent assays (ELISAs) have been widely employed in water testing for cyanotoxins. However, it still has issues, such as low sensitivity [11] or inaccuracy. Erroneous detection is due to the cross-reactivity of isomorphic substances with targets. False-positive results can occur in a worst-case scenario [12]. The high performance liquid chromatography (HPLC) and liquid chromatography-mass spectrometry (LC–MS) are the most accurate analytical methods and have been often employed in cyanotoxin testing [11, 13–16]. But they require exquisite instruments and complicated operations, making them not as affordable as ELISA-based testing. Aside from these limitations, chemical testing can only tell the presence and/or quantity of cyanotoxins without identifying the toxin producer(s). However, it is crucial to recognize the existence of toxigenic cyanobacteria in water bodies for monitoring and early warning of cyanotoxin poisoning incidents.

It is known that cyanotoxin synthesis is catalyzed by a string of relevant enzymes encoded by toxin synthetase genes [17–23]. Lack of essential genes for forming a toxin backbone or disruption of the enzymatic cascade toward toxin production results in the failure of toxin synthesis. Therefore, the detection of toxin synthetase genes in samples by a molecular test can disclose the presence or absence of toxigenic cyanobacteria. In this chapter, we review the application of molecular techniques, particularly PCR-based assays, for detecting toxigenic cyanobacteria in freshwater.

2. General genomic organization of toxigenic cyanobacteria

Like other bacteria, cyanobacteria often have one circular chromosome and a few plasmids that consist of the whole genome. The cyanobacterial chromosome is a few megabases in size and contains most of the genes, while plasmids play a role in transferring DNA elements. Compared to the eukaryotic microalgae, the cyanobacterial genome is highly compressed but still contains all genes essential for aquatic and photosynthetic life. Some species even have genes that can facilitate competitive superiority in the environment. For example, gas vesicle genes in *Planktothrix* spp. encode structural proteins that can form gas vesicles, endowing the cells with more buoyancy to the water's surface to gain more sunlight (i.e., solar energy) [24]. In addition, cyanotoxins were found effective in suppressing the growth of non-toxic species so that the toxigenic cyanobacteria have more survival advantages [25].

Cyanotoxin synthetase genes often cluster together in the genome and constitute one or more operons that are transcribed in identical or opposite directions [19, 21–23, 26]. The reason for such an arrangement is likely that the transcription can be well regulated so that all pertaining genes are transcribed simultaneously. This process may ensure that all necessary enzymes/proteins are present for subsequent toxin synthesis. The whole-genome sequencing of toxic cyanobacteria to date has demonstrated only a single copy of the toxin gene cluster in the cyanobacterial genome [27–29]. The toxin synthetase genes have conserved sequences encoding conserved domains/motifs in the corresponding proteins with specific functions during toxin syntheses, such as polyketide synthesis, adenylation, and methylation. The genes are always clustered closely with whose proteins conduct successive functions in a cascade reaction. It should be reiterated that the synthetase genes are indispensable for toxin production, making them the ideal targets for molecular detection.

Cyanotoxins are traditionally named after the first identified toxin-producing genus, as in the case of microcystin (*Microcystis*), anatoxin-a (*Anabaena*), cylindrospermopsin (*Cylindrospermopsis*), and so on. However, many different genera can produce the same cyanotoxin, indicative of the fact that these intergeneric toxic species have similar genetic elements for toxin production. For example, microcystin and microcystin synthetase genes (*mcy*) are reportedly found in *Microcystis*, *Anabaena*, *Planktothrix*, and *Aphanizomenon* [30]. Nevertheless, the gene clusters are disparate between genera regarding sequences, gene numbers, constitutions, and relative loci [17, 20, 26, 31, 32]. Such a characteristic is believed to be caused by divergent evolution from the common ancestors [33] or horizontal gene transfer [34]. Therefore, PCR identification of toxigenic cyanobacteria is usually designed at the genus level, although there have been reports of detecting multiple genera producing the same cyanotoxin based on the conserved intergeneric sequences [35].

3. Cyanotoxins and toxin biosynthesis

3.1 Microcystin

Microcystin is the most common cyanotoxin implicated in human and animal poisoning incidents [36–38]. It is a hepatotoxin and thus can cause severe impairment in the liver when ingested by the casualties. The toxin is known to be produced by several genera of cyanobacteria, such as *Microcystis*, *Anabaena*, and *Planktothrix*, to name a few.

Microcystin is a cyclic heptapeptide that inhibits the eukaryotic protein phosphatase type 1 and 2A in humans and animals by forming an irreversible covalent bond to a cysteine in the catalytic domain of these enzymes. It consists of the following amino acids: D-alanine, X, D-MeAsp (D-erythro-ß-methyl-aspartic acid), Z, Adda ((2S,3S,8S,9S)-3-amino-9-methoxy-2,6,8-trimethyl-10-phenyldeca-4,6-dienoic acid), D-glutamic acid, and Mdha (N-methyldehydroalanine). X and Z represent variable L amino acids. It has reportedly over 80 variants, mostly differing in amino acids at the positions X and Z [39].

Microcystin is a non-ribosomal oligopeptide, which means unlike most of the peptides and proteins, it is not synthesized by cellular ribosomes. The enzymes responsible for its synthesis contain the non-ribosomal peptide synthetases (NRPS) and polyketide synthases (PKS) modules as well as tailoring functional domains. All the enzymes are the protein products encoded by the microcystin synthetase genes (*mcy*) that cluster together in the genome (**Table 1**, **Figure 1**). In *Microcystis*, ten *mcy* genes (*mcyA-J*) span 55 kb near the center of its 5.8 Mb circular chromosome and form two operons (*mcyABC* and *mcyD-J*) of which the transcription proceeds in discretely opposite directions [26, 40]. The 55.4 kb *mcyA-J* gene cluster in *Anabaena* also form two operons (*mcyABC* and *mcyG-DJEFIH*), one of which the gene order differs from *Microcystis* [31]. In contrast, *Planktothrix* has a 55.6 kb *mcy* cluster including eight essential genes (*mcyABC-DEGHJ*) that form a single operon and one unique gene (*mcyT*), and the arrangement and sequence of specific domains in the gene products differ from those in other genera [17].

Per annotation of *mcy* genes, microcystin biosynthesis is initiated by McyG to covalently bind a phenylacetate precursor that is then methylated by McyJ. Next, McyD elongates the growing chain by accepting a malonyl-CoA, and McyE introduces another malonyl-CoA and further extends the backbone of microcystin. As a racemase, McyF is involved in either the supply of D-glutamate or D-MeAsp, or the peptidyl epimerization of L-glutamate. Then McyA captures an L-serine and installs it into the growing chain, and the Mdha moiety is synthesized by McyI. Following the addition of an amino acid into position X and a D-MeAsp by McyB, McyC adds the last amino acid into position Z. Finally, mature microcystin is formed by the cyclization of the linear precursor. McyH, as an ATP-binding cassette (ABC) transporter, may be accountable for the transmembrane export of

Gene	Size (bp)[1]	Encoded domain or function[2]	Existence in different genera		
			Microcystis	*Anabaena*	*Planktothrix*
mcyA	8838	NRPS, C, NMT, E	yes	yes	yes
mcyB	6318	NRPS, A, T, C	yes	yes	yes
mcyC	3876	NRPS, C, A, TE	yes	yes	yes
mcyD	11721	PKS, KS, AT, KR, DH, ACP, CM	yes	yes	yes
mcyE	10464	PKS, NRPS, KS, AT, ACP, CM, AMT	yes	yes	yes
mcyF	756	Racemase	yes	yes	no
mcyG	7896	NRPS, PKS, KS, AT, CM, DH, KR, ACP	yes	yes	yes
mcyH	1617	Transporter	yes	yes	yes
mcyI	1014	Dehydrogenase	yes	yes	no
mcyJ	837	OM	yes	yes	yes
mcyT	< 1000	TE	no	no	yes

[1]*Values are from Microcystis aeruginosa PCC7806 [26] except mcyT from Planktothrix agardhii CYA 126 [17].*
[2]*NRPS, non-ribosomal peptide synthetase; C, condensation; NMT, N-methyltransferase; E, epimerization; T, thiolation; TE, thioesterase; PKS, polyketide synthase; KS, β-ketoacyl synthase; AT, acyltransferase; KR, β-ketoacyl reductase; DH, dehydratase; ACP, acyl carrier protein; CM, C-methyltransferase; AMT, aminotransferase; OM, O-methyltransferase.*

Table 1.
Comparison of microcystin synthetase genes.

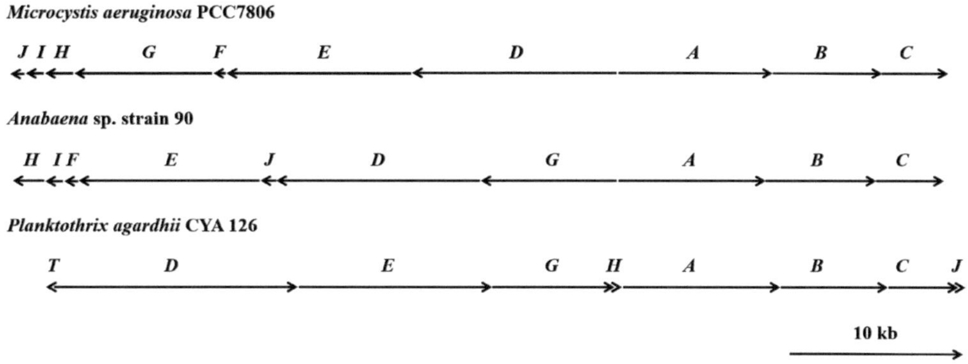

Figure 1.
The microcystin synthetase gene (mcy) clusters of Microcystis aeruginosa PCC7806, Anabaena sp. strain 90, and Planktothrix agardhii CYA 126. Each gene is indicated by its designated letter above the arrows, and genes shorter than 500 bp are drawn as an arrowhead only. The diagram only exhibits the size and position of each gene without the distances between two adjacent genes. The measuring scale arrow represents 10 kb.

microcystin. The *Planktothrix*-exclusive McyT is a putative thioesterase that may edit the synthesis by removing mis-primed amino acids from the NRPS and PKS enzymes.

3.2 Anatoxin-a

The cyanobacterial alkaloid anatoxin-a has been found in different genera, such as *Anabaena*, *Oscillatoria*, *Phormidium*, and *Cylindrospermum* [30, 41]. It is a neurotoxin that can bind to the neuronal nicotinic acetylcholine receptors and affect signal transmission between neurons and muscles as a nicotinic agonist [42]. By persistently stimulating the receptors to release signals for muscular contraction, the toxin eventually leads to respiratory arrest until victims' death in a few minutes. Homoanatoxin-a, in which a methyl group displaces a hydrogen atom at the end of the straight chain of anatoxin-a, is a natural analog of anatoxin-a and is also a potent nicotinic agonist.

Although anatoxin-a doesn't look structurally complicated, its synthesis still requires a cascade of many enzymes whose genes known as anatoxin-a synthetase genes (*ana*) have distinctive arrangements and sequences across genera like *mcy* genes (**Table 2**, **Figure 2**). For example, the toxic *Anabaena* has two clusters (20.3 kb *anaBCDEFG* and 1.7 kb *anaIAJ*) located 6.9 kb apart and transcribed in head-to-head opposite directions. The first cluster contains two operons (*anaBCDEF* and *anaG*) with a 288 bp gap between the operons [19, 32]. In contrast, the *ana* genes in *Oscillatoria* compose a ~ 23 kb cluster *anaJABCDEFGH* and a single upstream *anaI* gene transcribed in the opposite direction [32, 43]. *Cylindrospermum* has the most complicated arrangement of *ana* genes. The *anaIAHJ* and *anaBCDKEFG* are clustered together and are transcribed in oppositely separative directions. The *anaH*, however, is transcribed reversely in the small *anaIAHJ* cluster.

To start the anatoxin-a synthesis, AnaC activates and tethers the precursor proline to AnaD, which covalently combines with the proline. Then AnaB dehydrogenates the heterocyclic ring of proline to form a "C=N" double bond. AnaE introduces a carbonyl group into its connection with the heterocycle passed from AnaD. Then AnaJ catalyzes a cyclization step to form the characteristic bicyclic ring structure of anatoxin-a by connecting the heterocyclic ring with the backbone. At the same time, the growing chain is bound to the acyl carrier protein domain of AnaF. Finally, the bicyclic thioester is transferred to AnaG for chain extension

Gene	Size (bp)[1]	Encoded domain or function[2]	Existence in different genera		
			Anabaena	*Oscillatoria*	*Cylindrospermum*
anaA	750	TE	yes	yes	yes
anaB	1143	Proline-ACP oxidase	yes	yes	yes
anaC	1596	Proline adenylation	yes	yes	yes
anaD	273	Acyl carrier	yes	yes	yes
anaE	6438	PKS, KS, AT, DH, ER, KR, ACP	yes	yes	yes
anaF	5619	PKS, KS, AT, DH, KR, ACP	yes	yes	yes
anaG	4896	PKS, KS, AT, CM, ACP	yes	yes	yes
anaH	< 1000	Transposase	no	yes	yes
anaI	< 2000	Transporter	yes	yes	yes
anaJ	723	Cyclase	yes	yes	yes
anaK	<1000	Reductase	no	no	yes

[1]Values are from Anabaena sp. Strain 37 [19] except anaH from Oscillatoria sp. PCC 6506 [32] and anaK from Cylindrospermum stagnale PCC 7417 [41].
[2]TE, thioesterase; ACP, acyl carrier protein; PKS, polyketide synthase; KS, β-ketoacyl synthase; AT, acyltransferase; DH, dehydratase; ER, enoylreductase; KR, ketoreductase; CM, C-methyltransferase.

Table 2.
Comparison of anatoxin-a synthetase genes.

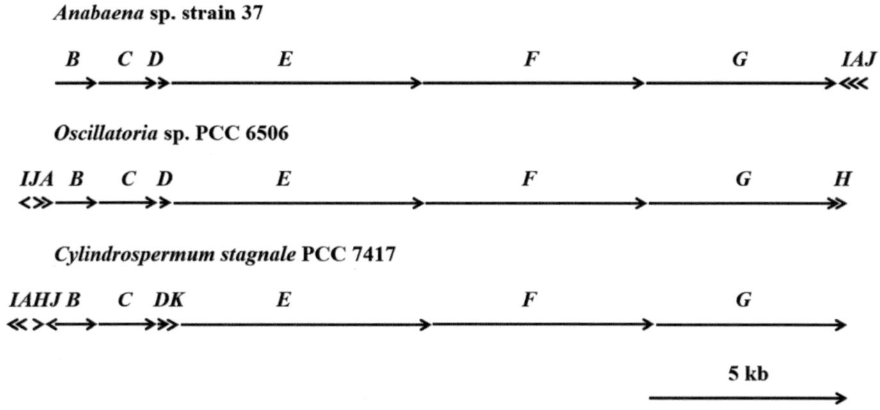

Figure 2.
The anatoxin-a synthetase gene (ana) clusters of Anabaena sp. strain 37, Oscillatoria sp. PCC 6506, and Cylindrospermum stagnale PCC 7417. Each gene is indicated by its designated letter above the arrows, and genes shorter than 250 bp are drawn as an arrowhead only. The diagram only exhibits the size and position of each gene without the distances between two adjacent genes. The measuring scale arrow represents 5 kb.

by adding an acyl group, followed by the enzymatic reaction of AnaA to break the single "SCO-C" covalent bond connecting the enzyme (AnaG) and final product for the completion and releasing of anatoxin-a. Similar to its counterpart McyH in microcystin-producing cyanobacteria, AnaI transports the toxin through the cytomembrane. The rest of the Ana proteins are not commonly shared across different genera and have their own functions. AnaH is a transposase only found in *Oscillatoria* and *Cylindrospermum* instead of *Anabaena*, implicating the toxic cyanobacteria in the former two genera were endowed with the toxin genes by intergeneric replicative transposition mechanism. *Cylindrospermum* has a unique AnaK that functions in further modification of anatoxin-a into dihydroanatoxin-a. The whole process is inferred as per the functional annotation of *ana* genes.

3.3 Cylindrospermopsin

Cylindrospermopsin can be produced by various cyanobacterial genera, such as *Cylindrospermopsis*, *Aphanizomenon*, *Anabaena*, and *Oscillatoria* [20, 30]. It is a cyclic sulfated guanidine alkaloid and can lead to cytotoxic, hepatotoxic, and neurotoxic impacts. Its molecule contains a central functional guanidino moiety, a hydroxymethyluracil ring, and a hetero tricyclic ring. The mechanism of its toxicity lies in many aspects, including the inhibition of glutathione and protein synthesis, the inhibition of cytochrome P450, and direct interaction with DNA [44–47].

Cylindrospermopsin is synthesized via a string of NRPS/PKS reactions conducted by up to over a dozen Cyr proteins (**Table 3**, **Figure 3**). The cylindrospermopsin synthetase genes (*cyr*) are clustered together but are also distinctive with respect to numbers, sequences, and organization in different genera. The toxic *Cylindrospermopsis* has the cluster *cyrDFGIKLHJABECMNO* in which *cyrK*, *cyrM*, and *cyrN* are transcribed in the opposite direction of the rest genes [22]. The *cyr* cluster in toxic *Aphanizomenon* and *Oscillatoria* is *cyrJMKIGFDCABE* and *cyrDC-orf1-cyrH-trp1-cyrKAB-trp2-cyrEJFGIN*, respectively [20, 48].

Gene	Size (bp)[1]	Encoded domain or function[2]	Existence in different genera		
			Cylindrospermopsis	*Aphanizomenon*	*Oscillatoria*
cyrA	1176	AMT	yes	yes	yes
cyrB	8754	NRPS, PKS, PCP, KS, AT, DH, MT, KR, ACP	yes	yes	yes
cyrC	5005	PKS, KS, AT, KR, ACP	yes	yes	yes
cyrD	5631	PKS, KS, AT, DH, KR, ACP	yes	yes	yes
cyrE	5667	PKS, KS, AT, DH, KR, ACP	yes	yes	yes
cyrF	4074	PKS, KS, AT, ACP	yes	yes	yes
cyrG	1437	Uracil ring formation	yes	yes	yes
cyrH	1431	Uracil ring formation	yes	yes	yes
cyrI	831	Hydroxylation	yes	yes	yes
cyrJ	780	Sulfotransferase	yes	yes	yes
cyrK	1398	Exporter	yes	yes	yes
cyrL	750	Transposase	yes	no	no
cyrM	318	Transposase	yes	no	no
cyrN	600	Adenylylsulfate kinase	yes	no	yes
cyrO	1548	Regulator	yes	no	yes
orf1	152	ATP-grasp protein	no	no	yes
trp1	404	Transposase	no	no	yes
trp2	299	Transposase	no	no	yes

[1]*Values are from Cylindrospermopsis raciborskii AWT205 [22] except orf1, trp1, and trp2 from Oscillatoria sp. Strain PCC 6506 [20].*
[2]*AMT, aminotransferase; NRPS, non-ribosomal peptide synthetase; PKS, polyketide synthase; PCP, peptidyl carrier protein; AT, acyltransferase; DH, dehydratase; MT, methyl transferase; KR, ketoreductase; ACP, acyl carrier protein; KS, β-ketoacyl synthase.*

Table 3.
Comparison of cylindrospermopsin synthetase genes.

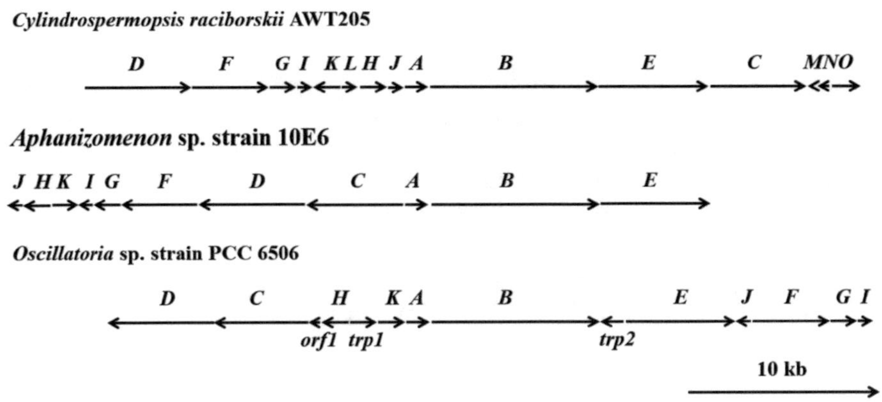

Figure 3.
The cylindrospermopsin synthetase gene (cyr) clusters of Cylindrospermopsis raciborskii AWT205, Aphanizomenon sp. strain 10E6, and Oscillatoria sp. strain PCC 6506. Each gene is indicated by its designated letter above the arrows, and genes shorter than 500 bp are drawn as an arrowhead only. The diagram only exhibits the size and position of each gene without the distances between two adjacent genes. The measuring scale arrow represents 10 kb.

As *cyr* genes have been annotated with exact biological functions, the cylindrospermopsin biosynthesis is inferred as follows. CyrA conducts the transamidination of an L-arginine to a glycine to form a guanidinoacetate. Then the product is activated by CyrB, and the first ring of the tricyclic structure is formed in the end. CyrC elongates the growing chain by an acetate via activation of a malonyl-CoA and its condensation with the chain. CyrD and CyrE catalyze the formation of the second and third rings of the tricyclic rings. CyrF accepts and extends the growing chain by adding an acetate. Next, CyrG and CyrH carry out the formation of the uracil ring, and CyrJ and CyrN together catalyze the sulfation at the hydroxyl group in the hetero tricyclic ring. At last, CyrI completes cylindrospermopsin synthesis by introducing a hydroxyl group to the carbon atom between the rings. In addition, CyrK is an exporter and resembles the function of McyH and AnaI. CyrO has diverse regulatory and signal transduction roles. The transposases CyrL and CyrM are exclusively found in *Cylindrospermopsis*. In comparison, *trp1* and *trp2* genes encode unique transposases in *Oscillatoria*, and the *orf1* gene, that is unique in *Oscillatoria*, codes for an ATP-grasp protein.

3.4 Nodularin

Nodularin is a cyclic pentapeptide and has the identical chemical structure as microcystin except the lack of D-alanine and the amino acid at position X. The mechanism of its toxicity is the same as microcystin's, i.e., inhibiting the eukaryotic protein phosphatase catalytic subunit type 1 and 2A and leading to severe liver damage. Different from the three aforementioned cyanotoxins, nodularin is solely found in *Nodularia* and synthesized by the nodularin synthetase gene (*nda*) cluster that contains two operons, *ndaAB* and *ndaDEFGHI*, transcribed in opposite directions (**Table 4**, **Figure 4**) [23]. Because of the remarkable resemblance with *mcy* clusters with a few missing regions and genes, *nda* genes could be regarded as the degenerative *mcy* genes.

Nodularin synthesis is conducted putatively according to the annotated functions of each Nda protein. NdaC activates the starter unit as phenylalanine or phenylacetate, and then NdaE catalyzes the transfer of a methyl group to the growing chain. NdaD is involved in two further polyketide extension steps, and NdaF facilitates the final round of polyketide extension and the biosynthesis of

Gene	Size (bp)[1]	Encoded domain or function[2]
ndaA	2607	NRPS, A, NM, PCP, C
ndaB	1299	C, A, PCP, TE
ndaC	2640	NRPS, PKS, A, PCP, KS, AT, CM, KR, ACP
ndaD	3872	PKS, KS, AT, CM, DH, KR, ACP
ndaE	927	OM
ndaF	3475	PKS, NRPS, KS, AT, CM, ACP, AMT, C, A, PCP
ndaG	235	Racemase
ndaH	341	D-3-phosphoglycerate dehydrogenase
ndaI	601	ABC transporter

[1]*Values are from Nodularia spumigena strain NSOR10 [23].*
[2]*NRPS, non-ribosomal peptide synthetase; A, adenylation; NM, N-methyltransferase; PCP, peptidyl carrier protein; C, condensation; TE, thioesterase; PKS, polyketide synthase; KS, ketosynthase; AT, acyltransferase; CM, C-methyltransferase; KR, ketoreductase; ACP, acyl carrier protein; DH, dehydratase; OM, O-methyltransferase; AMT, aminotransferase.*

Table 4.
Comparison of nodularin synthetase genes.

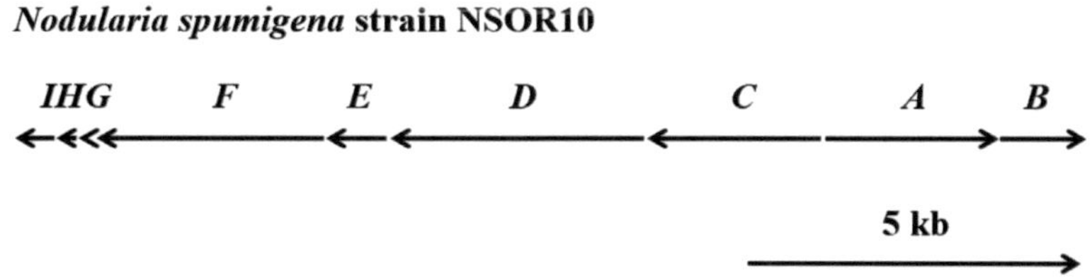

Figure 4.
The nodularin synthetase gene (nda) clusters of Nodularia spumigena strain NSOR10. Each gene is indicated by its designated letter above the arrows, and genes shorter than 250 bp are drawn as an arrowhead only. The diagram only exhibits the size and position of each gene without the distances between two adjacent genes. The measuring scale arrow represents 5 kb.

Adda. Next, epimerization of L-glutamic acid is catalyzed by NdaG, followed by the peptide condensation carried out by NdaA and NdaB. During the condensation, NdaH participates in the conversion of N-methyl-L-threonine (MeThr) to N-methyldehydrobutyrine (MeDhb) with a cofactor nicotinamide adenine dinucleotide (NADH). Finally, the mature peptide chain is cyclized by NdaB and released from the enzyme-substrate complex. As an ABC-transporter, NdaI is responsible for the transmembrane transportation of nodularin for extracellular excretion.

4. PCR detection of toxic cyanobacteria

PCR-based assays have been most commonly utilized in molecular identification studies because the assays are able to recognize targets accurately. The assays incorporate oligonucleotide primers explicitly designed for complementary sequences of the target gene(s). Two types of PCR methods have been used: conventional gel-based PCR and real-time PCR. In general, the real-time PCR has higher sensitivity (i.e., detect a low amount of the target) than the conventional PCR. The real-time PCR also offers better specificity than the conventional PCR since it uses an additional oligonucleotide known as a probe, which is complementary to sequences between primer-binding sequences.

Furthermore, the real-time PCR allows estimating the number of the intended target in samples when performed with standards with a known copy number of the target sequences. This procedure is referred to as quantitative real-time PCR (qPCR). In addition, reverse transcription (RT)-PCR or RT-qPCR platforms have been utilized for specifically detecting transcripts (i.e., mRNAs) from the target genes of cyanobacteria. Typically, PCR can be completed within one or two hours, much shorter than the traditional analytical methods and microscopy mentioned above.

4.1 Microcystin-producing cyanobacteria

The molecular identification of microcystin-producing cyanobacteria has been conducted using nearly all *mcy* genes; nonetheless, most studies have selected *mcyA*, *B*, *D*, and *E* as the target genes for *Microcystis*. Tillett et al. designed PCR primers from the *N*-methyltransferase domain of *mcyA* gene of *Microcystis*, evaluated those primers on 37 *Microcystis* strains with and without toxin production and found the molecular outcomes were significantly in concordance with the toxicity of each strain [49]. Kurmayer et al. designed *Microcystis*-specific primers based on *mcyB* gene and observed that the proportion of toxic cells in the overall *Microcystis* population correlated positively with the size of *Microcystis* colonies [50]. Using *mcyD* as an indicator, Kaebernick et al. reported that light had a positive effect on the transcriptional response of *mcy* gene cluster in *Microcystis* over specific threshold intensities [51].

Although most publications have been concerned about toxic/toxigenic *Microcystis*, there are reports of identifying other toxigenic genera with their *mcy* genes. Toxic *Planktothrix* in a French lake was identified and quantified by qPCR using primers devised in the condensation and adenylation domains of *mcyA* gene but was accounted for only 54% of the variation in microcystin levels [52]. Mbedi et al. reported the use of a highly variable region in *mcyE* encoding the adenylation domain to design primers specific to *Planktothrix*, which were validated by 46 *Planktothrix* strains in a conventional PCR assay [53]. Vaitomaa et al. developed two *mcyE*-based qPCR assays specific to *Anabaena* and *Microcystis*, respectively, and utilized them to investigate the two toxic genera in two Finnish lakes. They concluded the microcystin concentrations correlated positively with the sum of *Anabaena* and *Microcystis mcyE* gene copy numbers [54]. Ngwa et al. also applied *mcyE*-based primers to specific detection of *Planktothrix* and *Microcystis*, respectively, using the qPCR and RT-qPCR assays and found a significant positive correlation between microcystin concentrations and abundances of *mcyE* genes rather than transcripts from the *mcyE* genes [55].

The rest of the *mcy* genes have been less often used for molecular detection in comparison to the four genes above. Yuan et al. developed a conventional PCR method for detecting toxigenic *Microcystis* based on *mcyC*. They demonstrated a good correlation between the presence of this gene and microcystin in water samples from farm ponds [56]. Mbedi et al. and Ouahid et al. used *mcyG* with a few other *mcy* genes as the targets for the recognition of toxigenic *Planktothrix* and *Microcystis*, respectively. They performed a multiplex PCR using the primers targeting *mcyG* and *mcyD* on field colonies and showed the same outcome as lab cultures, suggesting that simultaneous amplification of several gene regions was feasible [53, 57]. *mcyJ* has been used for the quantification of microcystin-producing *Microcystis* genotypes via qPCR in ecological investigations in China and Korea. Interestingly, Zhang et al. reported a weak correlation of gene numbers versus toxin concentrations, whereas Joung et al. found a strong correlation [58–60]. Although *mcyT* was chosen in the specific detection of toxigenic *Planktothrix* due to its uniqueness, the gene was also found in non-toxigenic strains in the same study, negating its candidacy for specific molecular detection of toxigenic *Planktothrix* [53]. No reports of using *mcyF* and *mcyI* as a target for a molecular assay have been made yet. These genes are not common in all *mcy* clusters disclosed to date, thus not good candidates for the detection purpose to cover a wide range of microcystin-producing cyanobacteria. As *mcyH* codes for a transporter that is not necessary for microcystin biosynthesis, the gene has not been generally considered as a suitable target for the identification of toxigenic cyanobacteria.

With increased bioinformatic data related to *mcy* genes in multiple cyanobacterial genera, the molecular identification of microcystin-producing cyanobacteria has proceeded to multi-genera detection. Conserved domains in *mcy* genes provide adequate genetic information for searching out consensus sequences across different toxic genera for multi-generic molecular identification. Hisbergues et al. designed a pair of primers from the condensation domain of *mcyA* that could facilitate the detection of microcystin-producing cyanobacteria including *Anabaena*, *Microcystis*, and *Planktothrix* by PCR, and the toxin producer could be recognized at the genus level by combining PCR with the restriction fragment length polymorphism assay [61]. Hautala et al. found consensus sequences in *mcyB* among *Anabaena*, *Microcystis*, and *Planktothrix*, devised specific primers and genus-specific probes for qPCR assays and demonstrated the positive correlation between gene copy numbers and toxin concentrations [62]. On the contrary, Ye et al. employed these primers in a survey of the cyanobacterial population producing microcystin to assess their dynamics and concentrations in a lake in China and found no correlation between gene copies and toxin concentrations [63]. Beversdorf et al. selected *mcyE* and *mcyA* as the genes of interest for designing primers specific for *Microcystis*, *Planktothrix*, and *Anabaena* in their qPCR assay and concluded *mcy* genes were not a good indicator of microcystin in the environment [35].

There are a few unidentified open reading frames (ORFs) flanking the *mcy* cluster, which may have relevant functions in microcystin synthesis, such as the *dnaN* and *uma* in *Microcystis* [26]. Nevertheless, these ORFs have not been used as targets for detecting the toxigenic cyanobacteria, probably due to their undetermined roles. The only example is that Tillet et al. obtained the corresponding amplicons by a PCR using *uma1* primers from 20 *Microcystis* strains and found the physical distances in the genome were consistent between *uma1* and *mcyC* across all strains [49].

4.2 Anatoxin-a-producing cyanobacteria

The *ana* genes have been widely used for genus-specific detection of anatoxin-a producing cyanobacteria via PCR. For *Anabaena*, Legrand et al. set up a nested-PCR assay based on *anaC* gene for detecting the toxigenic planktonic *Dolichospermum* (previously known as *Anabaena*) and averted possible none-amplification in a few anatoxin-a-producing strains due to less stringent specificity [64]. For *Aphanizomenon*, Ballot et al. confirmed the discovery of an anatoxin-a-producing strain in a German lake by detecting the production of the toxin from and the existence of *anaF* in the cyanobacteria using a PCR with the primers designed out of the gene [65]. For *Oscillatoria*, the *anaG* region encoding the methylation domain was amplified and sequenced for the strains producing anatoxin-a and homoanatoxin-a isolated from a French river [66]. For *Phormidium*, Wood et al. detected *anaF* gene in 20 strains isolated from two rivers in New Zealand tested positive for anatoxin-a, homoanatoxin-a, dihydroanatoxin-a, or dihydrohomoanatoxin-a, disclosing a total agreement between the presence of toxins and the existence of genes [67]. However, such an agreement was not observed in all toxigenic strains when Rantala-Ylinen et al. chose *anaC* as the gene of interest and designed genus-specific primers for *Oscillatoria* and *Anabaena*, respectively [19].

Like *mcy* gene cluster, *ana* has also been used for multi-generic detection for anatoxin-a-producing cyanobacteria by PCR. For instance, Rantala-Ylinen et al. designed *anaC* primers specific for three toxigenic genera, *Anabaena*, *Oscillatoria*, and *Aphanizomenon*. They revealed the presence of both *Anabaena* and *Oscillatoria* as potential anatoxin-a producers in Finnish freshwaters and the Baltic Sea [19].

4.3 Cylindrospermopsin-producing cyanobacteria

Molecular detection of cylindrospermopsin-producing cyanobacteria has been mostly reported for *Cylindrospermopsis*. Burford et al. established a qPCR method based on the *cyrA* gene to detect the toxigenic *Cylindrospermopsis* in field blooms in an Australian reservoir. They found the increase in cell quotas (i.e., toxin amount per cell) of cylindrospermopsin correlated with the increase in the proportion of *cyrA*/16S rDNA in the blooms [68]. Moreira et al. directly applied the K18/M4 primer set that Fergusson et al. [69] designed based on *cyrC* to evaluate toxigenic *Cylindrospermopsis* abundance and toxicological potential by qPCR in a lake in Portugal. They found only one out of ten samples were positive for *cyrC* and cylindrospermopsin [70]. Another pair of *cyrC* primers cyl2/cyl14 was designed to recognize toxigenic *Cylindrospermopsis* via PCR by Wilson et al. [71]. That primer set was utilized by Marbun et al. in the on-site monitoring of the toxic cyanobacteria in reservoirs in Taiwan using qPCR [72]. The authors revealed good accordance between cylindrospermopsin concentrations and toxigenic *Cylindrospermopsis* cell numbers.

Multi-generic detection of cylindrospermopsin-producing cyanobacteria was reported as well. Campo et al. found that *cyrJ* is the gene suitable for designing primers and probes and established a Taqman qPCR assay for specific detection of toxigenic *Aphanizomenon* and *Cylindrospermopsis*. The presence of the *cyrJ* gene in cyanobacteria was in concordance with the toxin production, as revealed by testing 11 experimental strains [73]. Fergusson et al. designed primers out of *cyrC* regions encoding polyketide synthase and peptide synthetase and combined them into a multiplex PCR. The PCR was able to identify the cylindrospermopsin-producing *Cylindrospermopsis*, *Anabaena*, and *Aphanizomenon*. In their study, the complete matching of positive/negative detection of gene versus toxin was shown by testing of 39 related strains [69].

There are also a few ORFs flanking the *cyr* cluster that encode proteins related to cylindrospermopsin synthesis, as shown with microcystin. Nevertheless, these ORFs have not been evaluated in detecting the toxic cyanobacteria due to their unidentified or unnecessary roles in toxin biosynthesis.

4.4 Nodularin-producing cyanobacteria

Since *nda* gene clusters are only found in *Nodularia*, all molecular identification studies were developed for this genus. Kruger et al. devised 11 pairs of primers for all nine *nda* genes in a comparative PCR study with toxigenic and non-toxigenic *Nodularia* strains and discovered that the lack of toxicity was caused by the absence of all the *nda* genes [74]. Koskenniemi et al. aimed at *ndaF* gene for primer design and set up a qPCR method for the detection of toxigenic *Nodularia* spp. in the Baltic Sea. A significant positive correlation was found between *ndaF* gene copy numbers and nodularin concentrations, referring to a relatively constant toxin production [75]. To investigate the expression of *nda* genes in a bloom-forming *Nodularia* strain from the Baltic Sea, Jonasson et al. designed nine pairs of primers for the nine *nda* genes and used them in an RT-qPCR assay. They observed that all genes were continuously expressed during growth. Still, the intracellular and extracellular nodularin concentrations did not vary significantly in contrast to the shifts in gene expression, indicating unknown regulatory mechanisms acting on the enzyme activity level and regulating the biosynthesis and/or the maturation of nodularin [76].

5. Other cyanotoxins and PCR detection of the toxic cyanobacteria

Apart from the four most commonly reported cyanotoxins mentioned above, there are a few other cyanotoxins, such as saxitoxin, lyngbyatoxin, guanitoxin, β-N-methylamino-L-alanin (BMAA), aplysiatoxin, and lipopolysaccharide [18, 77, 78]. Hitherto, only the gene clusters for the biosynthesis of saxitoxin and lyngbyatoxin have been characterized.

Saxitoxin belongs to the group of carbamate alkaloid toxins composed of a tetrahydropurine group and two guanidinium moieties [79] and can also be produced by marine phytoplankton [80]. It can cause paralytic shellfish poisoning syndrome and afflict human health via bioaccumulation. At least 30 clustered saxitoxin synthesis genes (*sxt*) have been reported to be involved in the biosynthesis of saxitoxin, which might be the most complicated within all known cyanotoxins [81]. Saxitoxin production has been found in multiple cyanobacteria genera, such as *Cylindrospermopsis*, *Anabaena*, *Aphanizomenon*, and *Lyngbya*, putatively due to frequent horizontal gene transfer [21, 34].

The *sxtA* gene of *Anabaena* was used as the template for primer designing in a qPCR assay. Still, the amplicon was also produced from three other saxitoxin-producing genera, demonstrating its multi-generic detection capacity [82]. The study also revealed that the saxitoxin concentrations correlated positively with *stx* gene copy numbers, indicating the latter can be used as a measure of potential toxigenicity in *Anabaena* and other cyanobacterial blooms. Al-Tebrineh et al. employed the primers by aligning *sxtA* and *cyrA* genes from four genera in both conventional PCR and qPCR for a field survey along an Australian river [83]. They found cyanobacteria with the genes were widespread and massive in the surveyed areas. The authors also suggested that the molecular method may be used as a proxy for bloom risk assessment due to the positive correlation between concentrations of each cyanotoxin and respective toxin gene copy numbers.

Lyngbyatoxin is characterized as a potent skin irritant produced by *Lyngbya* which has been found in estuarine and coastal waters in tropical and subtropical regions [18]. Its biosynthesis reportedly involves four lyngbyatoxin synthetase genes (*ltxA-D*). However, molecular detection of lyngbyatoxin-producing cyanobacteria using these genes has not been documented in the public domain to date.

No literature regarding molecular detection of cyanobacteria producing the rest of the toxins mentioned above could be searched. It is most likely because there are few reports as to the molecular mechanisms of their biosynthesis. Nevertheless, it is worthwhile to briefly introduce guanitoxin, previously known as anatoxin-a(S), to emphasize its difference from anatoxin-a. Guanitoxin was recently renamed due to its structural and toxicological disparities from anatoxin-a [77]. It is a guanidino organophosphate neurotoxin that irreversibly inhibits acetylcholinesterase's active site, leading to excess acetylcholine, which causes severe salvation and chromodacryorrhea, so-called "bloody tears" before respiratory arrest [84]. Up to now, it was only found in planktonic *Dolichospermum* that was previously designated as *Anabaena*.

6. Perspectives

As various cyanobacterial genera can produce the same cyanotoxin, the development of toxigenic cyanobacteria identification needs to be multi-generic detection. Furthermore, as many genes for different toxins have sequences for the same conserved domains, designing PCR methods for all the cyanobacteria producing multiple toxins would be ideal.

Although most publications have focused on the selected cyanotoxins and their producers, more attention should be paid to other cyanotoxins and producers due to their potential of posing a significant threat to animal and human health. However, many cyanotoxin-producing cyanobacteria still lack bioinformation for the synthesis-related genes (e.g., guanitoxin), and it is thereby urgent to make further exploration to enrich the gene pools and their sequences so that a much more comprehensive understanding of the molecular mechanisms and the development of nucleic acid-based identification methods can be facilitated.

With the technical advance in PCR, researchers have been able to develop multiplex PCR methods in which many cyanotoxin biosynthesis genes can be detected simultaneously. For example, Ouahid et al. devised a multiplex PCR assay to detect six *mcy* genes (*mcyA-E*, *G*) at the same time for the typing of toxigenic *Microcystis* [85]. Rasmussen et al. developed a duplex qPCR for the detection of toxigenic *Cylindrospermopsis* using *cyrA*, *cyrB*, and *cyrC* genes [86]. Al-Tebrineh et al. established a quadruplex qPCR assay that could concurrently detect *mcyE*, *ndaF*, *cyrA*, and *sxtA* genes for most of the toxin-producing genera in a single reaction [87] and successfully employed the assay in an investigation of a cyanobacterial bloom that occurred along a river in Australia [83]. The invention of multiplex PCR assays has enhanced the throughput significantly for toxigenic cyanobacteria detection and can provide great aid to large-scale ecological surveys.

Cyanobacteria with cyanotoxin synthetase genes in their genome are clearly equipped with the ability of toxin production. However, transcription of toxin biosynthesis genes is triggered by various environmental factors [88–90]; hence, toxin production is not consistently ongoing. It means the presence of genes itself may not always translate into the appearance of toxins unless they are inter- or extra-cellularly accumulated and detectable. Furthermore, the significant positive correlation between gene copies and toxin levels is still controversial, as described in this chapter and another review [91]. Instead, the presence of mRNA transcripts from cyanotoxin synthase genes may be more closely associated with toxin production. Consequently, cDNA detection is justifiable to indicate an ongoing toxin synthesis, which is more critical and useful for monitoring the toxin-producing cyanobacteria. For this purpose, genes located at the end of operons should be good candidates for two reasons. One, primers designed from those genes can be directly used in cDNA testing like other genes because cyanobacteria lack introns. Two, the appearance of those genes in cDNA form signifies the successful cascade transcription of the clustered genes, gearing up all pertinent proteins for toxin synthesis. For example, *mcyC* and *mcyJ* located at the 3′ terminals of each operon in toxigenic *Microcystis* would be more useful in this respect than *mcyA*, *B*, *D*, and *E* which are more often used. However, it should be noted that a significant positive correlation might not exist between cDNA copies and toxin levels, likely due to the incomplete transcription or complex regulation of transcription and the elusive fate of toxins [55].

Although qPCR is preferred due to its many advantages, conventional PCR should also be considered for assessing the presence or absence of toxigenic cyanobacteria in water samples, as previously reported [49, 56]. In addition, the simplicity and cheaper operation may make conventional PCRs a more cost-effective tool for molecular detection of toxigenic cyanobacteria in comparison to qPCRs.

Besides PCR-based assays, there are other molecular technologies applicable to the identification and/or characterization of toxigenic cyanobacteria. A noteworthy method is the next-generation sequencing (NGS) technology. The technology has been widely used to identify previously unrecognized agents, non-culturable microorganisms, and/or variants because of its advanced and hypothesis-free sequencing ability [92] and has been applied to cyanobacteria research. Although most NGS studies have been investigations of taxonomic diversities using representative cyanobacterial genetic markers such as 16S rDNA [93, 94], the potential toxigenicity of cyanobacteria can be disclosed by sequencing the pooled libraries of toxin biosynthesis associated genes. Casero et al. revealed the existence of multiple toxigenic taxa in a summer bloom in a Spanish reservoir using *mcyE*, *anaF*, and *stxI* genes, and the relative abundance of toxigenic cyanobacteria in those populations correlated with the respective toxin concentrations [94]. Another method worthwhile to mention is DNA microarray/chip technology which has also been employed in the identification of toxigenic cyanobacteria. For example, four microcystin-producing genera were detected by a genus-specific DNA chip assay based on *mcyE* gene by Rantala et al. [95]. These high-throughput technologies can serve the purpose of molecular identification of toxigenic cyanobacteria in conjunction with PCR.

7. Conclusions

Nowadays, freshwater cyanobacterial blooms are seen more frequently than ever before because of increased eutrophication of their habitats and climate changes (e.g., global warming), which are utterly favorable to the overgrowth of cyanobacteria. Even though toxic cyanobacterial species are not always the mere culprit for these ecological disasters, they are often the dominant organisms and cause more destructive consequences because they can produce potent cyanotoxins into the water. There is no doubt that the toxic freshwater cyanobacteria pose a grave threat to human and animal health, agricultural production, tourism, to name a few. Hence, advancing techniques and technologies for rapid and reliable identification and monitoring of toxic cyanobacteria is an inevitable mission for healthcare, economy, and environmental conservation. To date, molecular assays, especially PCR-based tests, have been employed in toxic cyanobacterial identification, but their utilization should be further expanded into large-scale and long-term detection tasks and routine monitoring programs for not only the acute poisoning incidents but also the chronic impacts and preventative measures.

Acknowledgements

The authors disclosed receipt of the following financial support for the publication of this chapter: the manuscript compilation was supported in part by funding from the Innovative Swine Industry Enhancement Grant Program by the Iowa Attorney General's Office, Iowa State University (ISU) Health Research Initiative, and ISU Veterinary Diagnostic Laboratory Research Support Fund.

Conflict of interest

The authors declare no conflict of interest.

Thanks

The authors are sincerely grateful to the people who have made their contributions to the pertaining studies that are helpful for writing the chapter, including, but not limited to, Dr. Steve Ensley, Dr. Hyun-Joong Kim, Dr. Christopher Filstrup, Dr. Baoqing Guo, Dr. Paula Imerman, Dr. Grace Wilkinson, Dwayne Schrunk, and Amy Curtis.

Author details

Jian Yuan and Kyoung-Jin Yoon*
Department of Veterinary Diagnostic and Production Animal Medicine, Iowa State University, Ames, USA

*Address all correspondence to: kyoon@iastate.edu

References

[1] Domingos P, Rubim TK, Molica RJR, Azevedo SM, Carmichael WW. First report of microcystin production by picoplanktonic cyanobacteria isolated from a northeast Brazilian drinking water supply. Environ Toxicol. 1999;14:31-35. DOI: 10.1002/(SICI)1522-7278(199902)14:1<31::AID-TOX6>3.0.CO;2-B

[2] Falconer IR. An overview of problems caused by toxic blue–green algae (cyanobacteria) in drinking and recreational water. Environ Toxicol. 1999;14:5-12. DOI: 10.1002/(SICI)1522-7278(199902)14:1<5::AID-TOX3>3.0.CO;2-0

[3] Falconer IR, Burch MD, Steffensen DA, Choice M, Coverdale OR. Toxicity of the blue-green alga (cyanobacterium)*Microcystis aeruginosa* in drinking water to growing pigs, as an animal model for human injury and risk assessment. Environ Toxic Water. 1994;9:131-139. DOI: 10.1002/tox.2530090209

[4] Zaccaroni A, Scaravelli D. Toxicity of fresh water algal toxins to humans and animals. In: Evangelista V, Barsanti L, Frassanito AM, Passarelli V, Gualtieri P, editors. Algal Toxins: Nature, Occurrence, Effect and Detection. Dordrecht: Springer; 2008. p. 45-89. DOI: 10.1007/978-1-4020-8480-5_3

[5] Carmichael WW, Boyer GL. Health impacts from cyanobacteria harmful algae blooms: Implications for the North American Great Lakes. Harmful Algae. 2016;54:194-212. DOI: 10.1016/j.hal.2016.02.002

[6] Paerl HW, Otten TG. Harmful cyanobacterial blooms: causes, consequences, and controls. Microb Ecol. 2013;65(4):995-1010. doi: 10.1007/s00248-012-0159-y

[7] Miller MA, Kudela RM, Mekebri A, Crane D, Oates SC, Tinker MT, et al. Evidence for a novel marine harmful algal bloom: cyanotoxin (microcystin) transfer from land to sea otters. PLoS One. 2010;5:e12576. DOI: 10.1371/journal.pone.0012576.

[8] Backer LC, Landsberg JH, Miller M, Keel K, Taylor TK. Canine cyanotoxin poisonings in the United States (1920s-2012): review of suspected and confirmed cases from three data sources. Toxins. 2013;5:1597-1628. DOI: 10.3390/toxins5091597

[9] Dittmann E, Wiegand C. Cyanobacterial toxins--occurrence, biosynthesis and impact on human affairs. Mol Nutr Food Res. 2006;50:7-17. DOI: 10.1002/mnfr.200500162. PubMed PMID: 16304634

[10] Stewart I, Seawright AA, Shaw GR. Cyanobacterial poisoning in livestock, wild mammals and birds – an overview. In: Hudnell HK, editor. Cyanobacterial Harmful Algal Blooms: State of the Science and Research Needs. New York: Springer; 2008. p. 613-637. DOI: 10.1007/978-0-387-75865-7_28

[11] Mathys W, Surholt B. Analysis of microcystins in freshwater samples using high performance liquid chromatography and an enzyme-linked immunosorbent assay. Int J Hyg Envir Heal. 2004;207:601-605. DOI: 10.1078/1438-4639-00334

[12] Brown A, Foss A, Miller MA, Gibson Q. Detection of cyanotoxins (microcystins/nodularins) in livers from estuarine and coastal bottlenose dolphins (*Tursiops truncatus*) from Northeast Florida. Harmful Algae. 2018;76:22-34. DOI: 10.1016/j.hal.2018.04.011

[13] Melvin A. Determination of microcystin-LR in municipal water using HPLC-UV/Vis. In: Proceedings of the National Conference on

Undergraduate Research (2017 NCUR); 6-8 April 2017; Memphis

[14] Shamsollahi HR, Alimohammadi M, Nabizadeh R, Nazmara S, Mahvi AH. Measurement of microcystin -LR in water samples using improved HPLC method. Global Journal of Health Science. 2014;7:66-70. DOI: 10.5539/gjhs.v7n2p66

[15] Robillot C, Vinh J, Puiseux-Dao S, Hennion M-C. Hepatotoxin production kinetics of the cyanobacterium *Microcystis aeruginosa* PCC 7820, as determined by HPLC–mass spectrometry and protein phosphatase bioassay. Environ Sci Technol. 2000;34:3372-3378. DOI: 10.1021/es991294v

[16] Lu N, Sun J, Kong F, Zhang D. Statistical analysis of a HPLC method for microcystins determination in drinking water. In: Proceedings of the International Conference on Electric Technology and Civil Engineering (ICETCE); 22-24 April 2011; Lushan. New York: IEEE; 2011. p. 4469-4470

[17] Christiansen G, Fastner J, Erhard M, Borner T, Dittmann E. Microcystin biosynthesis in *Planktothrix*: genes, evolution, and manipulation. J Bacteriol. 2003;185:564-572. DOI: 10.1128/jb.185.2.564-572.2003

[18] Dittmann E, Fewer DP, Neilan BA. Cyanobacterial toxins: biosynthetic routes and evolutionary roots. FEMS Microbiol Rev. 2013;37:23-43. DOI: 10.1111/j.1574-6976.2012.12000.x

[19] Rantala-Ylinen A, Kana S, Wang H, Rouhiainen L, Wahlsten M, Rizzi E, et al. Anatoxin-a synthetase gene cluster of the cyanobacterium *Anabaena* sp. strain 37 and molecular methods to detect potential producers. Appl Enviro Microb. 2011;77:7271-7278. DOI: 10.1128/AEM.06022-11

[20] Mazmouz R, Chapuis-Hugon F, Mann S, Pichon V, Mejean A, Ploux O. Biosynthesis of cylindrospermopsin and 7-epicylindrospermopsin in *Oscillatoria* sp. strain PCC 6506: identification of the *cyr* gene cluster and toxin analysis. Appl Enviro Microb. 2010;76:4943-4949. DOI: 10.1128/AEM.00717-10

[21] Mihali TK, Kellmann R, Neilan BA. Characterisation of the paralytic shellfish toxin biosynthesis gene clusters in *Anabaena circinalis* AWQC131C and *Aphanizomenon* sp. NH-5. BMC Biochem. 2009;10:8. DOI: 10.1186/1471-2091-10-8

[22] Mihali TK, Kellmann R, Muenchhoff J, Barrow KD, Neilan BA. Characterization of the gene cluster responsible for cylindrospermopsin biosynthesis. Appl Enviro Microb. 2008;74:716-722. DOI: 10.1128/AEM.01988-07

[23] Moffitt MC, Neilan BA. Characterization of the nodularin synthetase gene cluster and proposed theory of the evolution of cyanobacterial hepatotoxins. Appl Enviro Microb. 2004;70:6353-6362. doi: 10.1128/AEM.70.11.6353-6362.2004

[24] Beard S, Davis P, Iglesias-Rodrıguez D, Skulberg O, Walsby A. Gas vesicle genes in *Planktothrix* spp. from Nordic lakes: strains with weak gas vesicles possess a longer variant of *gvpC*. Microbiology. 2000;146:2009-2018. DOI: 10.1099/00221287-146-8-2009

[25] Yang J, Deng X, Xian Q, Qian X, Li A. Allelopathic effect of *Microcystis aeruginosa* on *Microcystis wesenbergii*: microcystin-LR as a potential allelochemical. Hydrobiologia. 2014;727:65-73. DOI: 10.1007/s10750-013-1787-z

[26] Tillett D, Dittmann E, Erhard M, Dohren HV, Borner T, Neilan BA. Structural organization of microcystin biosynthesis in *Microcystis aeruginosa* PCC7806: an integrated

peptide–polyketide synthetase system. Chem Biol. 2000;7:753-764. DOI: 10.1016/S1074-5521(00)00021-1

[27] Brown NM, Mueller RS, Shepardson JW, Landry ZC, Morre JT, Maier CS, et al. Structural and functional analysis of the finished genome of the recently isolated toxic *Anabaena* sp. WA102. BMC Genomics. 2016;17:457. DOI: 10.1186/s12864-016-2738-7

[28] Wang H, Sivonen K, Rouhiainen L, Fewer DP, Lyra C, Rantala-Ylinen A, et al. Genome-derived insights into the biology of the hepatotoxic bloom-forming cyanobacterium *Anabaena* sp. strain 90. BMC Genomics. 2012;13:613. DOI: 10.1186/1471-2164-13-613

[29] Rounge TB, Rohrlack T, Nederbragt AJ, Kristensen T, Jakobsen KS. A genome-wide analysis of nonribosomal peptide synthetase gene clusters and their peptides in a *Planktothrix rubescens* strain. BMC Genomics. 2009;10:396. DOI: 10.1186/1471-2164-10-396

[30] Buratti FM, Manganelli M, Vichi S, Stefanelli M, Scardala S, Testai E, et al. Cyanotoxins: producing organisms, occurrence, toxicity, mechanism of action and human health toxicological risk evaluation. Arch Toxicol. 2017;91:1049-1130. DOI: 10.1007/s00204-016-1913-6

[31] Rouhiainen L, Vakkilainen T, Siemer BL, Buikema W, Haselkorn R, Sivonen K. Genes coding for hepatotoxic heptapeptides (microcystins) in the cyanobacterium *Anabaena* strain 90. Appl Enviro Microb. 2004;70:686-692. DOI: 10.1128/aem.70.2.686-692.2004

[32] Mejean A, Paci G, Gautier V, Ploux O. Biosynthesis of anatoxin-a and analogues (anatoxins) in cyanobacteria. Toxicon. 2014;91:15-22. DOI: 10.1016/j.toxicon.2014.07.016

[33] Rantala A, Fewer DP, Hisbergues M, Rouhiainen L, Vaitomaa J, Borner T, et al. Phylogenetic evidence for the early evolution of microcystin synthesis. Proc Natl Acad Sci. 2004;101:568-573. DOI: 10.1073/pnas.0304489101

[34] Kellmann R, Michali TK, Neilan BA. Identification of a saxitoxin biosynthesis gene with a history of frequent horizontal gene transfers. J Mol Evol. 2008;6:526-538. DOI: 10.1007/s00239-009-9210-0

[35] Beversdorf LJ, Chaston SD, Miller TR, McMahon KD. Microcystin *mcyA* and *mcyE* gene abundances are not appropriate indicators of microcystin concentrations in lakes. PLoS One. 2015;10:e0125353. DOI: 10.1371/journal.pone.0125353

[36] Pouria S, de Andrade A, Barbosa J, Cavalcanti RL, Barreto VTS, Ward CJ, et al. Fatal microcystin intoxication in haemodialysis unit in Caruaru, Brazil. Lancet. 1998;352:21-26. DOI: 10.1016/s0140-6736(97)12285-1

[37] Fitzgerald SD, Poppenga RH. Toxicosis due to microcystin hepatotoxins in three Holstein heifers. J Vet Diagn Invest. 1993;5:651-653. DOI: 10.1177/104063879300500433

[38] Classen DM, Schwartz KJ, Madson D, Ensley SM. Microcystin toxicosis in nursery pigs. J Swine Health Prod. 2017;25:198-205.

[39] Sivonen K, Jones G. Cyanobacterial toxins. In: Chorus I, Bartram J, editors. Toxic Cyanobacteria in Water: A Guide to Their Public Health Consequences, Monitoring and Management. Boca Raton; CRC Press; 1999. p. 43-112. DOI: 10.1201/9781482295061

[40] Kaneko T, Nakajima N, Okamoto S, Suzuki I, Tanabe Y, Tamaoki M, et al. Complete genomic structure of the bloom-forming toxic cyanobacterium *Microcystis aeruginosa* NIES-843. DNA

Res. 2007;14:247-256. DOI: 10.1093/dnares/dsm026

[41] Méjean A, Dalle K, Paci G, Bouchonnet S, Mann S, Pichon V, et al. Dihydroanatoxin-a is biosynthesized from proline in *Cylindrospermum stagnale* PCC 7417: isotopic incorporation experiments and mass spectrometry analysis. J Nat Prod. 2016;79:1775-1782. DOI: 10.1021/acs.jnatprod.6b00189

[42] Devlin J, Edwards O, Gorham P, Hunter N, Pike R, Stavric B. Anatoxin-a, a toxic alkaloid from *Anabaena flos-aquae* NRC-44h. Can J Chem. 1977;55:1367-1371. DOI: 10.1139/v77-189

[43] Méjean A, Mann S, Maldiney T, Vassiliadis G, Lequin O, Ploux O. Evidence that biosynthesis of the neurotoxic alkaloids anatoxin-a and homoanatoxin-a in the cyanobacterium *Oscillatoria* PCC 6506 occurs on a modular polyketide synthase initiated by L-proline. J Am Chem Soc. 2009;131:7512-7513. DOI: 10.1021/ja9024353

[44] Runnegar MT, Kong S-M, Zhong Y-Z, Lu SC. Inhibition of reduced glutathione synthesis by cyanobacterial alkaloid cylindrospermopsin in cultured rat hepatocytes. Biochem Pharmacol. 1995;49:219-225. DOI: 10.1016/S0006-2952(94)00466-8

[45] Runnegar MT, Xie C, Snider BB, Wallace GA, Weinreb SM, Kuhlenkamp J. In vitro hepatotoxicity of the cyanobacterial alkaloid cylindrospermopsin and related synthetic analogues. Toxicol Sci. 2002;67:81-87. DOI: 10.1093/toxsci/67.1.81

[46] Humpage AR, Fenech M, Thomas P, Falconer IR. Micronucleus induction and chromosome loss in transformed human white cells indicate clastogenic and aneugenic action of the cyanobacterial toxin, cylindrospermopsin. Mutat Res-Gen Tox En 2000;472:155-161. DOI: 10.1016/S1383-5718(00)00144-3

[47] Kiss T, Vehovszky A, Hiripi L, Kovacs A, Vörös L. Membrane effects of toxins isolated from a cyanobacterium, *Cylindrospermopsis raciborskii*, on identified molluscan neurones. Comp Biochem Phys C. 2002;131:167-176. DOI: 10.1016/S1532-0456(01)00290-3

[48] Stuken A, Jakobsen KS. The cylindrospermopsin gene cluster of *Aphanizomenon* sp. strain 10E6: organization and recombination. Microbiology. 2010;156:2438-2351. DOI: 10.1099/mic.0.036988-0

[49] Tillett D, Parker DL, Neilan BA. Detection of toxigenicity by a probe for the microcystin synthetase A gene (*mcyA*) of the cyanobacterial genus *Microcystis*: comparison of toxicities with 16S rRNA and phycocyanin operon (Phycocyanin Intergenic Spacer) phylogenies. Appl Environ Microb. 2001;67:2810-2818. DOI: 10.1128/AEM.67.6.2810-2818.2001

[50] Kurmayer R, Christiansen G, Chorus I. The abundance of microcystin-producing genotypes correlates positively with colony size in *Microcystis* sp. and determines its microcystin net production in Lake Wannsee. Appl Environ Microb. 2003;69:787-95. DOI: 10.1128/aem.69.2.787-795.2003

[51] Kaebernick M, Neilan BA, Börner T, Dittmann E. Light and the transcriptional response of the microcystin biosynthesis gene cluster. Appl Environ Microb. 2000;66:3387-3392. DOI: 10.1128/AEM.66.8.3387-3392.2000

[52] Briand E, Gugger M, Francois JC, Bernard C, Humbert JF, Quiblier C. Temporal variations in the dynamics of potentially microcystin-producing

strains in a bloom-forming *Planktothrix agardhii* (Cyanobacterium) population. Appl Environ Microb. 2008;74:3839-3848. DOI: 10.1128/AEM.02343-07

[53] Mbedi S, Welker M, Fastner J, Wiedner C. Variability of the microcystin synthetase gene cluster in the genus *Planktothrix* (Oscillatoriales, Cyanobacteria). FEMS Microbiol Lett. 2005;245:299-306. DOI: 10.1016/j.femsle.2005.03.020

[54] Vaitomaa J, Rantala A, Halinen K, Rouhiainen L, Tallberg P, Mokelke L, et al. Quantitative real-time PCR for determination of microcystin synthetase E copy numbers for *Microcystis* and *Anabaena* in lakes. Appl Environ Microb. 2003;69:7289-7297. DOI: 10.1128/aem.69.12.7289-7297.2003

[55] Ngwa FF, Madramootoo CA, Jabaji S. Comparison of cyanobacterial microcystin synthetase (*mcy*) E gene transcript levels, *mcyE* gene copies, and biomass as indicators of microcystin risk under laboratory and field conditions. MicrobiologyOpen. 2014;3(4):411-25. doi: 10.1002/mbo3.173

[56] Yuan J, Kim H-J, Filstrup CT, Guo B, Imerman P, Ensley S, et al. Utility of a PCR-based method for rapid and specific detection of toxigenic *Microcystis* spp. in farm ponds. J Vet Diagn Invest. 2020;32:369-381. DOI: 10.1177/1040638720916156

[57] Ouahid Y, Perez-Silva G, del Campo FF. Identification of potentially toxic environmental *Microcystis* by individual and multiple PCR amplification of specific microcystin synthetase gene regions. Environ Toxicol. 2005;20:235-242. DOI: 10.1002/tox.20103

[58] Zhang W, Lou I, Ung WK, Kong Y, Mok KM. Analysis of cylindrospermopsin- and microcystin-producing genotypes and cyanotoxin concentrations in the Macau storage reservoir. Hydrobiologia. 2014;741:51-68. DOI: 10.1007/s10750-013-1776-2

[59] Kim S-G, Joung S-H, Ahn C-Y, Ko S-R, Boo SM, Oh H-M. Annual variation of *Microcystis* genotypes and their potential toxicity in water and sediment from a eutrophic reservoir. FEMS Microbiol Ecol. 2010;74:93-102. DOI: 10.1111/j.1574-6941.2010.00947.x

[60] Joung S-H, Oh H-M, Ko S-R, Ahn C-Y. Correlations between environmental factors and toxic and non-toxic *Microcystis* dynamics during bloom in Daechung Reservoir, Korea. Harmful Algae. 2011;1:188-193. DOI: 10.1016/j.hal.2010.09.005

[61] Hisbergues M, Christiansen G, Rouhiainen L, Sivonen K, Börner T. PCR-based identification of microcystin-producing genotypes of different cyanobacterial genera. Arch Microbiol. 2003;180:402-410. DOI: 10.1007/s00203-003-0605-9

[62] Hautala H, Lamminmaki U, Spoof L, Nybom S, Meriluoto J, Vehniainen M. Quantitative PCR detection and improved sample preparation of microcystin-producing *Anabaena*, *Microcystis* and *Planktothrix*. Ecotox Environ Safe. 2012;87:49-56. DOI: 10.1016/j.ecoenv.2012.10.008

[63] Ye W, Liu X, Tan J, Li D, Yang H. Diversity and dynamics of microcystin—producing cyanobacteria in China's third largest lake, Lake Taihu. Harmful Algae. 2009;8:637-644. DOI: 10.1016/j.hal.2008.10.010

[64] Legrand B, Lesobre J, Colombet J, Latour D, Sabart M. Molecular tools to detect anatoxin-a genes in aquatic ecosystems: Toward a new nested PCR-based method. Harmful Algae. 2016;58:16-22.

[65] Ballot A, Fastner J, Lentz M, Wiedner C. First report of anatoxin-a-producing cyanobacterium

Aphanizomenon issatschenkoi in northeastern Germany. Toxicon. 2010;56:964-971. DOI: 10.1016/j.toxicon.2010.06.021

[66] Mann S, Cohen M, Chapuis-Hugon F, Pichon V, Mazmouz R, Mejean A, et al. Synthesis, configuration assignment, and simultaneous quantification by liquid chromatography coupled to tandem mass spectrometry, of dihydroanatoxin-a and dihydrohomoanatoxin-a together with the parent toxins, in axenic cyanobacterial strains and in environmental samples. Toxicon. 2012;60:1404-1414. DOI: 10.1016/j.toxicon.2012.10.006

[67] Wood SA, Smith FM, Heath MW, Palfroy T, Gaw S, Young RG, et al. Within-mat variability in anatoxin-a and homoanatoxin-a production among benthic *Phormidium* (cyanobacteria) strains. Toxins. 2012;4:900-912. DOI: 10.3390/toxins4100900

[68] Burford MA, Davis TW, Orr PT, Sinha R, Willis A, Neilan BA. Nutrient-related changes in the toxicity of field blooms of the cyanobacterium, *Cylindrospermopsis raciborskii*. FEMS Microbiol Ecol. 2014;89:135-148. DOI: 10.1111/1574-6941.12341

[69] Fergusson KM, Saint CP. Multiplex PCR assay for *Cylindrospermopsis raciborskii* and cylindrospermopsin-producing cyanobacteria. Environ Toxicol. 2003;18:120-125. DOI: 10.1002/tox.10108

[70] Moreira C, Martins A, Azevedo J, Freitas M, Regueiras A, Vale M, et al. Application of real-time PCR in the assessment of the toxic cyanobacterium *Cylindrospermopsis raciborskii* abundance and toxicological potential. Appl Microbiol Biot. 2011;92:189-197. DOI: 10.1007/s00253-011-3360-x

[71] Wilson KM, Schembri MA, Baker PD, Saint CP. Molecular characterization of the toxic cyanobacterium *Cylindrospermopsis raciborskii* and design of a species-specific PCR. Appl Environ Microb. 2000;66:332-338. DOI: 10.1128/AEM.66.1.332-338.2000

[72] Marbun YR, Yen HK, Lin T-F, Lin HL, Michinaka A. Rapid on-site monitoring of cylindrospermopsin-producers in reservoirs using quantitative PCR. Sustain Environ Res. 2012;22:143-151.

[73] Campo E, Agha R, Cirés S, Quesada A, El-Shehawy R. First TaqMan assay to identify and quantify the cylindrospermopsin-producing cyanobacterium *Aphanizomenon ovalisporum* in water. Adv Microbiol. 2013;3:430-437. DOI: 10.4236/aim.2013.35058

[74] Krüger T, Oelmüller R, Luckas B. Comparative PCR analysis of toxic *Nodularia spumigena* and non-toxic *Nodularia harveyana* (Nostocales, Cyanobacteria) with respect to the nodularin synthetase gene cluster. Eur J Phycol. 2009;44:291-295. DOI: 10.1080/09670260802588434

[75] Koskenniemi K, Lyra C, Rajaniemi-Wacklin P, Jokela J, Sivonen K. Quantitative real-time PCR detection of toxic *Nodularia* cyanobacteria in the Baltic Sea. Appl Environ Microb. 2007;73:2173-2179. DOI: 10.1128/AEM.02746-06

[76] Jonasson S, Vintila S, Sivonen K, El-Shehawy R. Expression of the nodularin synthetase genes in the Baltic Sea bloom-former cyanobacterium *Nodularia spumigena* strain AV1. FEMS Microbiol Ecol. 2008;65:31-39. DOI: 10.1111/j.1574-6941.2008.00499.x

[77] Fiore MF, de Lima ST, Carmichael WW, McKinnie SM, Chekan JR, Moore BS. Guanitoxin, re-naming a cyanobacterial organophosphate toxin. Harmful Algae.

2020;92:101737. DOI: 10.1016/j.hal.2019.101737

[78] Stewart I, Schluter PJ, Shaw GR. Cyanobacterial lipopolysaccharides and human health - a review. Environ Health. 2006;5:1-23. DOI: 10.1186/1476-069X-5-7

[79] Schantz EJ, Ghazarossian V, Schnoes HK, Strong F, Springer J, Pezzanite JO, et al. Structure of saxitoxin. J Am Chem Soc. 1975;97:1238-1239. DOI: 10.1021/ja00838a045

[80] Gu H. Morphology, phylogenetic position, and ecophysiology of *Alexandrium ostenfeldii* (Dinophyceae) from the Bohai Sea, China. J Syst Evol. 2011;49:606-616. DOI: 10.1111/j.1759-6831.2011.00160.x

[81] Kellmann R, Mihali TK, Jeon YJ, Pickford R, Pomati F, Neilan BA. Biosynthetic intermediate analysis and functional homology reveal a saxitoxin gene cluster in cyanobacteria. Appl Environ Microb. 2008;74:4044-4053. DOI: 10.1128/AEM.00353-08

[82] Al-Tebrineh J, Mihali TK, Pomati F, Neilan BA. Detection of saxitoxin-producing cyanobacteria and *Anabaena circinalis* in environmental water blooms by quantitative PCR. Appl Environ Microb. 2010;76:7836-7842. DOI: 10.1128/AEM.00174-10

[83] Al-Tebrineh J, Merrick C, Ryan D, Humpage A, Bowling L, Neilan BA. Community composition, toxigenicity, and environmental conditions during a cyanobacterial bloom occurring along 1,100 kilometers of the Murray River. Appl Environ Microb. 2012;78:263-272. DOI: 10.1128/AEM.05587-11

[84] Carmichael WW, Gorham PR. Anatoxins from clones of *Anabaena flos-aquae* isolated from lakes of western Canada. Internationale Vereinigung für Theoretische und Angewandte Limnologie: Mitteilungen. 1978;21:285-295. DOI: 10.1080/05384680.1978.11903972

[85] Ouahid Y, Del Campo FF. Typing of toxinogenic *Microcystis* from environmental samples by multiplex PCR. Appl Microbiol Biot. 2009;85:405. DOI: 10.1007/s00253-009-2249-4

[86] Rasmussen JP, Giglio S, Monis P, Campbell R, Saint C. Development and field testing of a real-time PCR assay for cylindrospermopsin-producing cyanobacteria. J Appl Microbiol. 2008;104:1503-1515. DOI: 10.1111/j.1365-2672.2007.03676.x

[87] Al-Tebrineh J, Pearson LA, Yasar SA, Neilan BA. A multiplex qPCR targeting hepato-and neurotoxigenic cyanobacteria of global significance. Harmful Algae. 2012;15:19-25. DOI: 10.1016/j.hal.2011.11.001

[88] Wood SA, Rueckert A, Hamilton DP, Cary SC, Dietrich DR. Switching toxin production on and off: intermittent microcystin synthesis in a *Microcystis* bloom. Env Microbiology Rep. 2011;3:118-124. DOI: 10.1111/j.1758-2229.2010.00196.x.

[89] Pimentel JSM, Giani A. Microcystin production and regulation under nutrient stress conditions in toxic *Microcystis* strains. Appl Environ Microb. 2014;80:5836-5843. DOI: 10.1128/aem.01009-14.

[90] Gobler CJ, Davis TW, Coyne KJ, Boyer GL. Interactive influences of nutrient loading, zooplankton grazing, and microcystin synthetase gene expression on cyanobacterial bloom dynamics in a eutrophic New York lake. Harmful Algae. 2007;6:119-133. DOI: 10.1016/j.hal.2006.08.003.

[91] Pacheco AB, Guedes IA, Azevedo SM. Is qPCR a reliable indicator of cyanotoxin risk in freshwater? Toxins. 2016;8:172. DOI: 10.3390/toxins8060172.

[92] Cao Y, Fanning S, Proos S, Jordan K, Srikumar S. A review on the application of next generation sequencing technologies as applied to food-related microbiome studies. Frontiers in Microbiology. 2017;8:1829. DOI: 10.3389/fmicb.2017.01829.

[93] Celikkol-Aydin S, Gaylarde CC, Lee T, Melchers RE, Witt DL, Beech IB. 16S Rrna gene profiling of planktonic and biofilm microbial populations in the Gulf of Guinea using Illumina NGS. Marine Environmental Research. 2016;122:105-112. DOI: 10.1016/j.marenvres.2016.10.001.

[94] Casero MC, Velázquez D, Medino-Cobo M, Quesada A, Cirés S. Unmasking the identity of toxigenic cyanobacteria driving a multi-toxin bloom by high-throughput sequencing of cyanotoxins genes and 16S rRNA metabarcoding. Science of the Total Environment. 2019;665:367-378. DOI: 10.1016/j.scitotenv.2019.02.083.

[95] Rantala A, Rizzi E, Castiglioni B, de Bellis G, Sivonen K. Identification of hepatotoxin-producing cyanobacteria by DNA-chip. Environmental Microbiology. 2008;10:653-664. DOI: 10.1111/j.1462-2920.2007.01488.x.

Chapter 2

Industrial Applications of Cyanobacteria

Ayesha Algade Amadu, Kweku Amoako Atta deGraft-Johnson and Gabriel Komla Ameka

Abstract

Cyanobacteria also known as blue-green algae are oxygenic photoautotrophs, which evolved ca. 3.5 billion years ago. Because cyanobacteria are rich sources of bioactive compounds, they have diverse industrial applications such as algaecides, antibacterial, antiviral and antifungal agents, hence, their wide use in the agricultural and health sectors. Cyanobacterial secondary metabolites are also important sources of enzymes, toxins, vitamins, and other pharmaceuticals. Polyhydroxyalkanoates (PHA) which accumulate intracellularly in some cyanobacteria species can be used in the production of bioplastics that have properties comparable to polypropylene and polyethylene. Some cyanobacteria are also employed in bioremediation as they are capable of oxidizing oil components and other complex organic compounds. There are many more possible industrial applications of cyanobacteria such as biofuel, biofertilizer, food, nutraceuticals, and pharmaceuticals. Additionally, the metabolic pathways that lead to the production of important cyanobacterial bioactive compounds are outlined in the chapter along with commercial products currently available on the market.

Keywords: agricultural applications, bioactive compounds, biofuels, bioplastics, cosmetics, cyanobacteria, environmental remediation, pharmaceuticals

1. Introduction

Cyanobacteria are one of the oldest organisms on earth, based on the evidence of 3.5 million year-old fossilized records [1]. These ancient organisms are ubiquitous photoautotrophic microorganisms found in fresh, brackish, marine or wastewater [2]. They are able to adapt to a wide range of conditions (salinities, temperatures, pH factors, light intensities, and so on). Oxygenic photosynthesis is believed to have started with these microbes, making them a big contributor to the oxygen-rich atmosphere we enjoy today. Cyanobacteria are by far, the only known prokaryotes that perform oxygen-evolving photosynthesis [3] and the concentration of CO_2 in the atmosphere would be twice as high had it not been for cyanobacteria [4]. Thus, cyanobacteria have helped to shape the evolutionary trajectory of the earth. It has been suggested that the oil-producing gene in plant chloroplasts originated from cyanobacteria. According to [5] a primordial plant cell "engulfed" a cyanobacterium about a billion years ago. The bacterium lived in the cell and supplied it with photosynthetic products. Thus, implying that the oil synthesis enzyme (acyltranferase) of the chloroplasts originated from cyanobacteria. Similarly, according to [6] ethylene production existed before land plants colonized the earth as evidenced by unambiguously homologous ethylene-signaling pathways in *Spirogyra* and *Arabidopsis*; implying that cell elongation was possibly an ancestral ethylene production response.

Nearly 2,700 species of cyanobacteria have been described, however, prediction models suggest that between 2,000 and 8,000 cyanobacteria species exist in nature [7]. Some of the described species have been studied and known to have potential applications in agriculture, energy, food and pharmaceutical industries because of their ability to produce oil, fix atmospheric nitrogen, and also have high vitamin, mineral and protein contents among others [8].

Eutrophic water bodies facilitate the proliferation of cyanobacteria cells into blooms. Some of the blooms consist of cyanotoxin-producing species. The cyanotoxins are poisonous to humankind, animals and aquatic life. This has led to the closure of recreational centres and caused huge economic losses [9]. It is, therefore, imperative to understand the actions of these metabolites, their biosynthesis and possible applications in industry. This review will discuss the mechanisms involved in the cellular production of important metabolites and address the far-reaching industrial applications of these cyanobacteria compounds. Commercial products and companies producing cyanobacteria products are highlighted.

2. Biology of cyanobacteria

The composition of the earth's earlier environment may have influenced the evolution of the metal resistance features and the metal-utilizing proteins in cyanobacteria during a period when the atmosphere was limited in oxygen. Thus, some cyanobacteria still thrive in low oxygen conditions and some strains are highly tolerant to free sulfide [10]. Additionally, cyanobacteria have a high tolerance to ultraviolet-B and -C radiations along with high temperature tolerance, as high as 73°C [10]. Their tolerance to radiation is likely to have been crucial in the early evolution of the cyanobacteria.

Morphologically, cyanobacteria cells are coccoid, filamentous (filamentous non-heterocyst forming, and heterocyst forming) forms [10, 11]. Cyanobacterial cells have an outermost peptidoglycan layer composed mainly of proteins and lipopolysaccharides. Other minor components of the peptidoglycan cell wall include carotenoids and lipids that enhance permeability, mechanical stability and resistance toward chemical substances [12]. Despite their overall gram-negative structure, the peptidoglycan layer in cyanobacteria is significantly thicker than that of most gram-negative bacteria [12]. Outside the cell wall is a carbohydrate-rich glycocalyx with varying proportion of three distinct layers; a closely associated sheath, a well-defined capsule, and loosely attached slime [13]. These three layers protect cyanobacteria cells from desiccation and possibly from predators and phages, making their applications in biotechnology more feasible due to their robust nature.

Cyanobacteria cells divide by fission (binary or multiple), fragmentation or spore-formation processes. The doubling time of cells can occur between 2.1 hours and 72 hours under optimal conditions and as long as 10,000 years under stress conditions such as *Chroococcidiopsis* in dry deserts of Antarctica [10]. Their presence in such dry conditions confirms their ability to tolerate desiccation and water stress, allowing their use in biological processes under arid conditions. The fast doubling time of cyanobacteria cells leads to rapid growth rate which makes their application in industry economically sustainable.

In addition to their fast growth rate, cyanobacteria possess superior photosynthetic capabilities that enables them to convert about 10% of the solar energy received, into biomass which is relative to the 1% conversion done by traditional energy crops like sugarcane or corn, grown for biofuel production [14]. Some cyanobacteria cells possess gas-filled cavities that allow them to float on water surfaces, enhancing light capture for better photosynthetic efficiency. Internal thylakoid membranes are the site of photosynthetic reactions in cyanobacteria. The presence of chlorophyll and phycobilins increase the capture and conversion of light energy during photosynthesis. Other pigments including xanthophylls, carotenes, c-phycoerythrin and c-phycocyanin are present in cyanobacteria. C-phycoerythrin and c-phycocyanin are unique to blue-green algae.

Circadian rhythms are very important in cyanobacteria. These rhythms are fundamental adaptations to the earth's daily light and temperature fluctuations that lead to proper metabolic activity [15]. For example, nitrogen fixation is common in many species of cyanobacteria. However, oxygenic photosynthesis and nitrogen fixation are discordant processes because the nitrogenase enzyme is inactivated by oxygen. Two mechanisms are used to separate these activities. First, a biological circadian clock separates them temporally and cellular differentiation separates them spatially. For instance, a unicellular species such as *Cyanothece* strain ATCC 51142 stores glycogen at daytime and fixes nitrogen at night [15], whereas the filamentous strain, *Trichodesmium erythraeum* IMS101 fixes nitrogen during the day in groups of specialized cells known as heterocysts [16]. Based on this, heterocyst-forming cyanobacteria are able to differentiate highly specialized cells to provide fixed nitrogen to the vegetative cells in a filament. This property allows cyanobacteria to be used as bio-fertilizers in agriculture because of their ability to fix atmospheric nitrogen [14].

Fluctuation in environmental conditions causes oxidative stress to cells which inevitably lead to increased reactive oxygen species (ROS) production. As a form of defense, photosynthetic microorganisms including cyanobacteria have developed several mechanisms to evade the negative effects of ROS. Antioxidants and polyunsaturated fatty acids (PUFAs) are known to provide protection to the cell against oxidative stress by stabilizing free radicals [17]. Among the many fatty acids, PUFAs are of great interest due to their numerous health benefits and increasing global demand [18]. Additionally, they help regulate various cellular processes such as oxygen and electron transport, membrane fluidity, and heat adaptation [19]. Cyanobacteria lack organelles such as endoplasmic reticulum, mitochondria, chloroplasts and Golgi apparatus typically found in eukaryotic cells. However, ribonucleic acid (RNA)-containing organelles known as ribosomes are widespread in cyanobacterial cells. Ribosomes are responsible for protein synthesis that enables cells to perform efficient metabolic activities.

3. Metabolic activities and the production of bioactive compounds

Cellular metabolic processes lead to the production of valuable primary and secondary metabolites. Photosynthesis and carotenogenesis are examples of metabolic processes that lead to the production of primary metabolites like lipids (e.g., PUFAs), antioxidants (e.g., carotenoids) and some proteins (e.g., primary proteins). Primary metabolites are directly involved in normal developmental processes such as cell division, growth and reproduction [20]; and can also be transformed into products such as bio-fertilizers, bio-plastics, nutrient supplements, and dyes. These are beneficial in industrial applications (**Figure 1**). Secondary

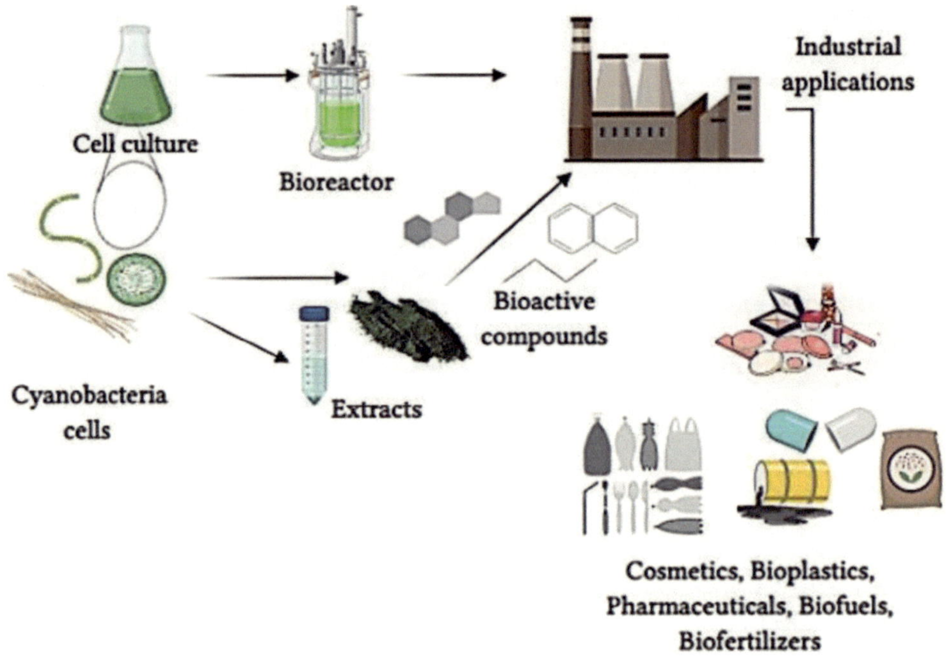

Figure 1.
Industrial applications of cyanobacteria from laboratory to up-scaled production and the extraction of metabolites of commercial value for the production of pharmaceuticals, biofertilizers, biofuels, biopolymers and cosmetics.

metabolites on the other hand, are not utilized by the cells for their primary needs. These include hormonal compounds and antibiotics, or toxins [21].

Cyanobacteria possess a relatively simple genome [14], making it easy for gene modification and manipulations for the exploration of novel metabolites. From a nutritional perspective, microbes can alter their cellular metabolism naturally through stress response such as nutrient deficiency [22]. Growth conditions can be manipulated to promote the production of biomass rich in valuable secondary metabolites of economic value. This is usually achieved through a two-stage culture technique where in the first step the cells are grown under optimal conditions to maximize biomass production. The second step involves the introduction of stress factors, such as nutrient deprivation or high light intensity, to induce the production of valuable secondary metabolites. Although a deficiency in nitrogen is known to inhibit cell cycle processes and the production of several cellular components, surprisingly, lipid and biopolymer (PHB) syntheses rather increase under these conditions [23]. This leads to the accumulation of oil droplets and starch granules in starved cells [22, 24]. These adaptive responses help to ensure the cell's survival during stressful conditions. Both metabolites (lipids and starch granules) serve as energy stores coupled with the special role of the starch granules in osmotic balance, heat, freezing, and ultraviolet (UV) rays [22, 23].

The methylerythritol 4-phosphate (MEP) pathway, known to occur in algae, bacteria, and plants, is a classic example of a metabolic pathway that is exploited for drug discoveries. In cyanobacteria, the MEP pathway leads to chlorophyll and hormone production [17]. It is observed that secondary metabolite production is species-specific and environmental conditions can influence their production. For instance, microalgae growing under stress conditions are more likely to produce secondary metabolites with antibacterial activity. In cyanobacteria, fatty acid synthesis (FAS) is performed by a type II fatty acid synthase complex and these fatty acids have anticarcinogenic, antibiotic, antifungal, and antiviral [25] properties that facilitates their use in pharmaceutical applications.

The shikimate pathway has been proposed to be responsible for mycosporine-like amino acids (MAAs) and scytonemin biosynthesis [26, 27]. These compounds have the potential to be used as natural UV blockers in product formulations such as cosmetic creams or paints, and varnishes (**Figure 1**).

4. Industrial applications of bioactive compounds

4.1 Bio-plastics

Cyanobacteria, have the potential to produce renewable biopolymers from natural resources such as, solar energy, water and CO_2, reducing the need for fertile soils, fertilizers, herbicides and potable water for crop production. Biodegradable polymers such as PHAs are produced as inclusion bodies within cells via the beta-oxidation pathway [28]. *Arthrospira (Spirulina)*, *Synechococcus*, and *Synechocystis*, are examples of cyanobacteria species widely employed in biopolymer production [23, 29]. Currently majority of commercial bioplastics produced from biologically-derived polymers are sourced from first generation feedstock through fermentation processes. Companies like **Tianjin GreenBio** (China), **Metabolix** (USA), **Biocycle** (Brazil) and **Polyferm** (Canada) produce biopolymers from sugars and vegetable oils. Some biopolymers are also produced by microorganisms like methanotrophs as in the case of **Mango Materials** (USA). Although the potential for biopolymer production from cyanobacteria exists, commercialization on an industrial scale is yet to be achieved. Research is currently geared towards optimization of cultivation and genetic modification approaches for enhanced biopolymer production [23, 30, 31].

4.2 Agricultural applications and environmental remediation

Cyanobacteria are known to play a key role in maintaining the stability of the surface crusts of dry lands. Thus, to combat desertification, cyanobacteria can be used in conjunction with bacteria, algae, mosses, lichens, or fungi, which form the biological soil crusts in unique geographical regions [32]. These biological soil crusts help primary succession in arid regions by improving the nutrient and moisture contents [32]. Cyanobacteria help to form a complex of heavy metals and xenobiotics to limit their mobility and transport in plants. Additionally, they offer protection to plants from disease-carrying insects, act as bio-control agents as well as enhance the mineralization of simpler organic molecules for easy assimilation. Additionally, cyanobacteria have been extensively applied in the area of environmental bioremediation through wastewater treatment processes [33, 34].

Arthrospira maxima and *A. platensis,* generically referred to as *Spirulina* are widely used as dietary supplements in feed for the poultry and aquaculture industries for their nutritional benefits due to their protein, vitamin and fatty acid content. Species such as *Anabaena, Aulosira, Calothrix, Nostoc,* and *Plectonema* are also used for agricultural purposes because of their nitrogen-fixing capabilities. They are able to fix atmospheric nitrogen (N_2) by the conversion of nitrogen to ammonia ($N_2 + 3H_2 \rightarrow 2NH_3$) also reducing soil salinity and controlling weed growth [35, 36]. Additionally, they increase soil phosphates [32] by converting insoluble phosphorus in the soil to phytoavailable forms [37], making them an excellent choice as bio-fertilizers. Studies have shown that endophytic cyanobacteria strains such as *Nostoc* strains have the ability to produce phytohormones, indole-3-acetic acid and cytokinins in root cells of both rice and wheat for their growth and development [38]. In Asian regions for instance, *Azolla* is either mixed into soil prior to rice planting or grown as a mixed crop along with rice in rice fields due to the presence of a cyanobacterium endosymbiont, *Anabaena azollae*, which fixes nitrogen [39].

Cyanobacteria also have other agriculture potential. Some cyanotoxins demonstrate biocidal activity. These biocides inhibit the growth of microorganisms such as viruses, bacteria and fungi; they also affect invertebrates including crustaceans, bivalves and vertebrates such as fish, birds, and mammals [40–44]. Thus, cyanobacteria toxins could be developed into active biological compounds and applied in crop fields as algaecides, fungicides, herbicides and insecticides because of their allelopathic effects [17, 43, 44]. Biocides have low environmental risks and are thus preferable to synthetic pesticides negatively affect the environment [17].

4.3 Biofuels

Our reliance on petroleum products has resulted in polluting the environment. Aside the release of toxic fumes including green-house gasses, into the atmosphere, the process of obtaining the fuel in itself presents potential environmental hazards. Cyanobacteria hold great promise as sources of renewable by-products (biodiesel) especially for the energy sector [45]. Cyanobacteria cells can be engineered to convert CO_2 and water into biofuels through photosynthesis. In a study [46] four modules were optimized to achieve high titer values (4.8 g/L^{-1}) of a petroleum substitute, 1-butanol. Firstly, 1-butanol biosynthesis was introduced and re-cast by systematic screening of genes and pathways. Module 2 involved the optimization of the 5′-regions of expression units to tune protein expression levels. Module 3 rewired carbon flux by editing acetate metabolism. In module 4, photosynthetic central carbon metabolism was rewritten by installing a phosphoketolase (PK) pathway. Several other biofuels have been produced from engineered cyanobacteria e.g., acetone, 2,3-butanediol ethanol, ethylene, isobutanol, 2-methyl-1-butanol from *Synechocystis sp.* [47]. **Algenol**, and **JouleUnlimited** are USA-based companies that use genetically modified cyanobacteria capable of growing in brackish or saltwater to produce a range of biofuels such as ethanol, biodiesel, gasoline and jet fuel in addition to other valuable chemicals [48–50]. Their ability to grow in saltwater makes their industrial application more economically and environmentally sustainable as it reduces the burden of using limited freshwater resources for their cultivation.

4.4 Cosmetics

Numerous cyanobacteria species have been used in the cosmetics industry for many decades because of their anti-inflammatory, antioxidant, and detoxifying properties [15, 17]. The cosmetic industry has evolved from just topical skin products to a more invasive approach of beautifying from within. This booming industry seeks to address skin-related issues by resolving internal problems at the cellular level. Skin aging, wrinkling, drying and other skin conditions occur as a result of loss of elasticity to the skin. In the cosmetic industry, for instance, a novel extract, extracellular polysaccharide (EPS), extracted from *Pseudomonas fluorescens* PGM37 has been found to have higher moisturizing retention ability and has the potential to be used in cosmetics and medicinal products [46]. Similarly, sacran a known cyanobacteria gel extracted from *Aphanothece sacrum* can be used as a moisturizing and anti-inflammatory agent [51, 52].

Again, microalgae strains with high amino acid content are great for improving skin texture and elasticity. Other strains rich in lipids help soothe and moisturize skin tissues, while antioxidant-rich algae with chlorophyll are ideal for detoxifying. Many Spirulina-infused cosmetic products are already on the market in the form of tablets, lotions and facial masks. Skin perfection has garnered so much attention in recent years. For instance, skin whitening has become a common practice world-wide, with a booming market in Asia and Africa [47]. Tyrosinase inhibition is the most common approach to achieving skin hypo-pigmentation as this enzyme catalyzes the rate-limiting step of pigmentation. Tyrosinase inhibitors have been isolated from numerous marine macroalgae species such as *Ecklonia cava*, *Laminaria japonica* and *Sargassum silquastrum* [47]. Oscillapeptin G, is an example of a tyrosinase inhibitor isolated from toxic cyanobacteria, *Oscillatoria agardhii* [48]. There is a huge potential for commercialization of cyanobacteria since their cells grow faster than that of marine macroalgae and thus would be more economically efficient at industrial scale.

4.5 Food

Food supplements, animal feed, food additives, and colorants produced from cyanobacterial carotenoids such as canthaxanthin, beta-carotene, nostoxanthin, and zeaxanthin are on the rise. *Spirulina* is a cyanobacterial specie rich in riboflavin, thiamine, β-carotene, and vitamin B_{12}. These supplements are sold as capsules, granules and tablets on the market. A carotenoid such as the ketocarotenoid, astaxanthin is known to be a more powerful antioxidant than vitamin C and A or other carotenoids which play a vital role in preventing damage in human cells through photooxidation [17]. Astaxanthin obtained from *Haematococcus pluvialis* contains protease inhibitors that may be used to treat diseases, such as HIV [17]. These food supplements are usually made from cyanobacterial biomass and consumed whole unlike extracts used in the production of pharmaceuticals [53].

4.6 Pharmaceuticals

Natural products have become important contributing sources of semi-synthetic and synthetic drugs in all major disease fields; predominantly in antibiotic therapies, immunoregulation and oncology [54–56]. Most of these bio-medical natural products or metabolites have been derived from cyanobacteria. These cyanobacterial metabolites have exhibited both interesting and exciting biological activities including antibacterial, anticancer, antifungal, antimicrobial, and antiviral activities. Others are anticoagulant, anti-HIV, anti-inflammatory, anti-malarial, antiprotozoal, antituberculosis, antitumor, and immunosuppressant activities [54, 55]. Some of these bioactive compounds are Borophycin from *Nostoc* sp. against human carcinoma, Calothrix from *Calothrix* sp. against human HELa cancer cells and inhibition of growth of chloroquine-resistant strain of malaria parasite *Plasmodium falciparum* [8, 54]. Extracts from *Lyngbya lagerhaimanni* has anti-HIV activity, whilst *Lyngbyatoxin* A from toxic strains of *Lyngbya majuscula* is highly inflammatory [56, 57].

4.7 Bio-pigments

Phycobilisomes are phycobiliproteins accumulated by cyanobacteria and these include phycocyanin (blue), phycoerythrin (red), and allophycocyanin (blue-gray). They are major light-harvesting complexes in cyanobacteria [58, 59]. These compounds are used as bio-pigments in industrial applications. *Linablue®*

is an example of such bio-pigments approved by the USFDA [60] for use in food products. After extraction of phycocyanin the waste biomass is used as the starting material for the production of black ink, *Algae Black*. In the past, charcoal and carbon-derived soot were the ingredients required to make black ink. These petrochemically-derived materials generated a significant carbon footprint, and listed as a class 2b carcinogen [61]. Currently, however, a product extracted from *Spirulina* is used as the base material to create the renewable, bio-based carbon black ink known as *AlgaeBlack* by the **Living Ink** company whose aim is to create a sustainable alternative to traditional carbon-black ink pigment. A summary of major cyanobacteria product-related companies along with their founding years are provided in **Table 1.** Chlorophylls for instance have been used as a textile dye with antimicrobial properties and also as a biomordant to enhance dyeing processes in textile production [8].

4.8 Research and development

Other applications of cyanobacterial extracts include their use in scientific research experiments. For instance, phycobiliproteins have fluorescent properties that can be used in flow cytometry and in immunoassay techniques [61]. Among these numerous bioactive compounds are terpenoids such as terpenes, diterpenes

Company/Country (founding year)	Focus area	Industrial applications	Reference
LG Sonic/Netherlands (1999)	Algal blooms	Effective ultrasonic frequencies for algae control on large water surface areas	[62]
Cyano Biotech/Germany (2004)	Development of novel cyanobacteria products	Discovery and development of novel structures based on cyanobacterial natural products	[63]
Photanol/Netherlands (2008)	Genetic modification of cyanobacteria	Produces broad range of biochemicals from cyanobacteria	[64]
Algae Biotechnologia/Brazil (2009)	Technological development of microalgae and cyanobacteria cultivation systems	Treatment of liquid and gaseous agro-industrial effluents. Production of ingredients and additives for animal nutrition. Production of human food supplements and production of biofuels.	[65]
Algenuity/United Kingdom (2009)	Synthetic biotechnology	Produces lab scale photobioreactor for algae and cyanobacteria research and also harness specific microalgal strains for synthetic biology applications	[66]
Living Ink Technology/USA (2013)	Production of ecofriendly ink products	Develops a variety of ink products and colors, including digital ink	[67]
Spira Inc./USA (2016)	Incorporation of algae-based ingredients into everyday products	Extract high-value compounds from algae.	[68]

Table 1.
Major industrial companies involved in the production of cyanobacterial metabolites.

and sesquiterpenes. These organic compounds are widely found in cyanobacteria and are used as natural ingredients in flavors and perfumes. Such applications are recently gaining grounds in therapeutic and pesticide industries [17]. Terpenoid type compounds such as carotenoids and phytols are crucial for chlorophyll and hormone biosynthesis through the methylerythritol 4-phosphate (MEP) pathway in cyanobacteria that was discussed in Section 3.

The rapid development of molecular tools for whole genome sequences promotes the use of omics technology i.e. transcriptomics, proteomics, and system biology approaches to manipulate metabolic pathways for producing valuable products [8].

Experimental approaches that require fluorescent probes such as fluorescence microscopy for diagnostics and biomedical research utilizes the autofluorescent properties of cyanobacteria pigments such as phycobiliproteins. Among these, the most widely exploited fluorescent probe is phycoerythrin utilized in biomedical research [8].

5. Conclusions

Bioactive compounds synthesized by cyanobacteria are innumerable coupled with their vast industrial applications. More of these natural compounds are being discovered on a regular basis through research and development. There also exists an untapped pool of bioactive compounds that genetic engineering techniques can unfold. Current process engineering strategies in cyanobacteria research is centered on the regulation of metabolic pathways for the production of bioactive compounds. Such exploitations are feasible due to the small and simple genome of cyanobacteria species. Considering the fact that cyanobacteria are a promising feedstock for a circular economy, it is important to develop robust strains suitable for industrial applications.

The possibility of producing novel biopolymer blends, biofuel components, and pharmaceutical compounds that are capable of meeting the demands of a biotechnologically-driven society is vast. Thus, cyanobacteria will continue to play crucial roles in terms of health, energy, food, and many other aspects of our lives.

To achieve great strides in the cyanobacteria production sector, a synergistic approach should be adopted by cyanobacteria-related companies. Collaboratively, companies dealing in products such as biofuels, and bioplastics that involve extraction of compounds from cyanobacteria cell biomass, could supply their waste biomass to other companies that require the waste biomass as raw materials for production, such as algae ink or biochar for wastewater treatment. Thus, forming a biorefinery supply chain network that will essentially benefit both the producers and consumers from an economic and environmental point of view. Cyanobacteria industrial harmonization will ensure the valorization of production processes.

Conflict of interest

The authors declare no conflict of interest.

Author details

Ayesha Algade Amadu[1], Kweku Amoako Atta deGraft-Johnson[2] and Gabriel Komla Ameka[3]*

1 Council for Scientific and Industrial Research – Water Research Institute, Accra, Ghana

2 CSIR-College of Science and Technology (CCST), Accra, Ghana

3 Department of Plant and Environmental Biology, University of Ghana, Legon, Accra, Ghana

*Address all correspondence to: gameka@ug.edu.gh

References

[1] Schopf J., Packer BM. Early Archean (3.3-Billion to 3.5-Billion-Year-Old) Microfossils from Warrawoona Group, Australia. Science (80-). 1987;237:70-3.

[2] Costa JAV, Moreira JB, Lucas BF, Da Silva Braga V, Cassuriaga APA, De Morais MG. Recent advances and future perspectives of PHB production by cyanobacteria. Ind Biotechnol. 2018; 14(5):249-56.

[3] Rasmussen B, Fletcher IR, Brocks JJ, Kilburn MR. Reassessing the first appearance of eukaryotes and cyanobacteria. Nature. 2008;455(7216): 1101-4.

[4] ScienceDaily. Unexpected discovery: Blue-green algae produce oil [Internet]. 2020. [cited 2021 Jan 15]. Available from: https://www.sciencedaily.com/releases/2020/03/200305132125.htm#:~:text=Cyanobacteria -- colloquially also called blue-green algae -- can,by the University of Bonn.

[5] Aizouq M, Peisker H, Gutbrod K, Melzer M, Hölzl G, Dörmann P. Triacylglycerol and phytyl ester synthesis in Synechocystis sp. PCC6803. Proc Natl Acad Sci U S A. 2020;117(11):6216-22.

[6] Ju C, Van De Poel B, Cooper ED, Thierer JH, Gibbons TR, Delwiche CF, et al. Conservation of ethylene as a plant hormone over 450 million years of evolution. Nat Plants [Internet]. 2015;1(1):1-7. Available from: http://dx.doi.org/10.1038/nplants.2014.4

[7] Nabout JC, da Silva Rocha B, Carneiro FM, Sant'Anna CL. How many species of Cyanobacteria are there? Using a discovery curve to predict the species number. Biodivers Conserv. 2013;22(12):2907-18.

[8] Singh PK, Kumar A, Kumar SV, Shrivastava AK, editors. Advances in Cyanobacterial Biology. Academic Press Inc.; 2020.

[9] WHO. Algae and cyanobacteria in fresh water. In: Guidelines for safe recreational water environments. 1999. p. 136-58.

[10] Whitton BA, Potts M. Introduction to the cyanobacteria. Vol. 9789400738, Ecology of Cyanobacteria II: Their Diversity in Space and Time. 2012. p. 1-13.

[11] Whitton BA, Potts M. The Ecology of Cyanobacteria [Internet]. Acta Universitatis Agriculturae et Silviculturae Mendelianae Brunensis. Kluwer Academic Publishers; 2000. Available from: http://publications.lib.chalmers.se/records/fulltext/245180/245180.pdf%0Ahttps://hdl.handle.net/20.500.12380/245180%0Ahttp://dx.doi.org/10.1016/j.jsames.2011.03.003%0Ahttps://doi.org/10.1016/j.gr.2017.08.001%0Ahttp://dx.doi.org/10.1016/j.precamres.2014.12

[12] Hoiczyk E, Hansel A. Cyanobacterial cell walls: News from an unusual prokaryotic envelope. J Bacteriol. 2000;182(5):1191-9.

[13] Kumar K, Mella-Herrera RA, Golden JW. Cyanobacterial heterocysts. Cold Spring Harb Perspect Biol. 2010;2(4):1-19.

[14] Zahra Z, Choo DH, Lee H, Parveen A. Cyanobacteria: Review of current potentials and applications. Environ - MDPI. 2020;7(2).

[15] Toepel J, Welsh E, Summerfield TC, Pakrasi HB, Sherman LA. Differential transcriptional analysis of the cyanobacterium *Cyanothece* sp. strain ATCC 51142 during light-dark and continuous-light growth. J Bacteriol. 2008;190(11):3904-13.

[16] Sandh G, El-Shehawy R, Díez B, Bergman B. Temporal separation of cell division and diazotrophy in the marine diazotrophic cyanobacterium *Trichodesmium erythraeum* IMS101. FEMS Microbiol Lett. 2009;295(2): 281-8.

[17] Singh R, Parihar P, Singh M, Bajguz A, Kumar J, Singh S, et al. Uncovering potential applications of cyanobacteria and algal metabolites in biology, agriculture and medicine: Current status and future prospects. Front Microbiol. 2017;8(APR):1-37.

[18] Steinhoff FS, Karlberg M, Graeve M, Wulff A. Cyanobacteria in Scandinavian coastal waters - A potential source for biofuels and fatty acids? Algal Res [Internet]. 2014;5(1):42-51. Available from: http://dx.doi.org/10.1016/j.algal.2014.05.005

[19] Cardozo KHM, Guaratini T, Barros MP, Falcão VR, Tonon AP, Lopes NP, et al. Metabolites from algae with economical impact. Vol. 146, Comparative Biochemistry and Physiology - C Toxicology and Pharmacology. Elsevier Inc.; 2007. p. 60-78.

[20] Martínez-Francés E, Escudero-Oñate C. Cyanobacteria and Microalgae in the Production of Valuable Bioactive Compounds. In: IntechOpen [Internet]. 2018. p. 105-28. Available from: http://dx.doi.org/10.1039/C7RA00172J%0A https://www.intechopen.com/books/advanced-biometric-technologies/liveness-detection-in-biometrics%0Ahttp://dx.doi.org/10.1016/j.colsurfa.2011.12.014

[21] Carmichael WW. Cyanobacteria secondary metabolites—the cyanotoxins. J Appl Bacteriol. 1992;72(6):445-59.

[22] Chen B, Wan C, Mehmood MA, Chang JS, Bai F, Zhao X. Manipulating environmental stresses and stress tolerance of microalgae for enhanced production of lipids and value-added products–A review. Bioresour Technol [Internet]. 2017;244:1198-206. Available from: http://dx.doi.org/10.1016/j.biortech.2017.05.170

[23] Amadu AA, Qiu S, Ge S, Addico GND, Ameka GK, Yu Z, et al. A review of biopolymer (Poly-β-hydroxybutyrate) synthesis in microbes cultivated on wastewater. Sci Total Environ [Internet]. 2021;756:143729. Available from: https://doi.org/10.1016/j.scitotenv.2020.143729

[24] Ge S, Champagne P, Plaxton WC, Leite GB, Marazzi F. Microalgal cultivation with waste streams and metabolic constraints to triacyl glycerides accumulation for biofuel production. Biofuels, Bioprod Biorefining. 2017;11(4):325-43.

[25] El-Baz FK, El-Senousy WM, El-Sayed AB, Kamel MM. In vitro antiviral and antimicrobial activities of *Spirulina platensis* extract. J Appl Pharm Sci. 2013;3(12):52-6.

[26] Soule T, Palmer K, Gao Q, Potrafka RM, Stout V, Garcia-Pichel F. A comparative genomics approach to understanding the biosynthesis of the sunscreen scytonemin in cyanobacteria. BMC Genomics. 2009;10:1-10.

[27] Rosic NN. Mycosporine-like amino acids: Making the foundation for organic personalised sunscreens. Mar Drugs. 2019;17(11):1-17.

[28] Ranaivoarisoa TO, Singh R, Rengasamy K, Guzman MS, Bose A. Towards sustainable bioplastic production using the photoautotrophic bacterium *Rhodopseudomonas palustris* TIE-1. J Ind Microbiol Biotechnol [Internet]. 2019;46(9-10):1401-17. Available from: https://doi.org/10.1007/s10295-019-02165-7

[29] Kamravamanesh D, Lackner M, Herwig C. Bioprocess Engineering

Aspects of Sustainable Polyhydroxy alkanoate Production in Cyanobacteria. Bioengineering. 2018;5(4):111-29.

[30] Jiang XR, Wang H, Shen R, Chen GQ. Engineering the bacterial shapes for enhanced inclusion bodies accumulation. Metab Eng [Internet]. 2015;29:227-37. Available from: http://dx.doi.org/10.1016/j.ymben.2015.03.017

[31] Carpine R, Olivieri G, Hellingwerf KJ, Pollio A, Marzoc chella A. Industrial production of poly-β-hydroxybutyrate from CO2: Can cyanobacteria meet this challenge? Processes. 2020;8(3):1-23.

[32] Rossi F, Li H, Liu Y, De Philippis R. Cyanobacterial inoculation (cyanobacterisation): Perspectives for the development of a standardized multifunctional technology for soil fertilization and desertification reversal. Earth-Science Rev [Internet]. 2017;171:28-43. Available from: http://dx.doi.org/10.1016/j.earscirev.2017.05.006

[33] Singh JS, Kumar A, Rai AN, Singh DP. Cyanobacteria: A precious bio-resource in agriculture, ecosystem, and environmental sustainability. Front Microbiol. 2016;7(APR):1-19.

[34] Martins J, Peixe L, Vasconcelos VM. Unraveling Cyanobacteria Ecology in Wastewater Treatment Plants (WWTP). Microb Ecol. 2011;62(2):241-56.

[35] Jochum M, Moncayo LP, Jo Y. Microalgal cultivation for biofertilization in rice plants using a vertical semi-closed airlift photobioreactor. PLoS One. 2018;13(9):1-13.

[36] Shweta S, Samuel K. Study of Cyanobacteria as biofertilizer from the rice field. world J Pharm Res. 2015;4(3):1696-706.

[37] Sahu D, Priyadarshani I, Rath B. Cyanobacteria -As Potential Biofertilizer. CIBTech J Microbiol Jul-Sept Oct-Dec. 2012;1(3):2319-386720.

[38] Hussain A, Hasnain S. Phytostimulation and biofertilization in wheat by cyanobacteria. J Ind Microbiol Biotechnol. 2011;38(1):85-92.

[39] Bocchi S, Malgioglio A. Azolla-Anabaena as a Biofertilizer for Rice Paddy Fields in the Po Valley , a Temperate Rice Area in Northern Italy. 2010;2010.

[40] Misra S, Kushik D. Growth Promoting Substances of Cyanobacteria. II. Detection of Amino Acids, Sugars and Auxins. Proc Indian natn Sc Acad [Internet]. 1989;B55(5&6):499-504. Available from: http://www.insa.nic.in/writereaddata/UpLoadedFiles/PINSA/Vol55B_1989_5and6_Art29.pdf

[41] Schwartz RE, Hirsch CF, Sesin DF, Flor JE, Chartrain M, Fromtling RE, et al. Pharmaceuticals from cultured algae. J Ind Microbiol. 1990;5(2-3):113-23.

[42] Burja AM, Banaigs B, Abou-Mansour E, Grant Burgess J, Wright PC. Marine cyanobacteria - A prolific source of natural products. Tetrahedron. 2001;57(46):9347-77.

[43] Berry JP. Cyanobacterial Toxins as Allelochemicals with Potential Applications as Algaecides, Herbicides and Insecticides. Mar Drugs. 2008;6(2):117-46.

[44] Rastogi RP, Sinha RP. Biotechnological and industrial significance of cyanobacterial secondary metabolites. Biotechnol Adv [Internet]. 2009;27(4):521-39. Available from: http://dx.doi.org/10.1016/j.biotechadv.2009.04.009

[45] Doamekpor LK, Onwona-Agyeman R, Ameka GK. Bioenergy :

Biodiesel from Freshwater Cyanobacterium Occurring in Ghana Green Microalgae and a. West African J Appl Ecol. 2019;27(2):51-60.

[46] Liu X, Miao R, Lindberg P, Lindblad P. Modular engineering for efficient photosynthetic biosynthesis of 1-butanol from CO2 in cyanobacteria. Energy Environ Sci. 2019;12(9): 2765-77.

[47] Nozzi NE, Oliver JWK, Atsumi S. Cyanobacteria as a Platform for Biofuel Production. Front Bioeng Biotechnol. 2013;1(August).

[48] Mazard S, Penesyan A, Ostrowski M, Paulsen TI, Egan S. Tiny Microbes with a Big Impact : The Role of Cyanobacteria and Their Metabolites in Shaping Our Future. Mar Drugs. 2016;14(97):1-19.

[49] Algenol Biotech [Internet]. [cited 2021 Feb 14]. Available from: www.algenol.com

[50] Joule Unlimited [Internet]. [cited 2021 Feb 14]. Available from: http://www.jouleunlimited.com

[51] Morone J, Alfeus A, Vasconcelos V, Martins R. Revealing the potential of cyanobacteria in cosmetics and cosmeceuticals — A new bioactive approach. Algal Res [Internet]. 2019;41(May):101541. Available from: https://doi.org/10.1016/j.algal.2019.101541

[52] Okeyoshi K, Okajima MK, Kaneko T. The cyanobacterial polysaccharide sacran: characteristics, structures, and preparation of LC gels. Polym J [Internet]. 2021;53(1):81-91. Available from: http://dx.doi.org/10.1038/s41428-020-00426-2

[53] Chronakis IS, Madsen M. Algal proteins. In: Handbook of food proteins [Internet]. Woodhead Publishing Limited; 2011. p. 353-94. Available from: http://dx.doi.org/10.1533/9780857093639.353

[54] Zainuddin EN, Mundt S, Wegner U, Mentel R. Cyanobacteria a potential source of antiviral substances against influenza virus. Med Microbiol Immunol. 2003;191(3-4):181-2.

[55] Silva T, S. Salomon P, Hamerski L, Walter J, B. Menezes R, Siqueira JE, et al. Inhibitory effect of microalgae and cyanobacteria extracts on influenza virus replication and neuraminidase activity. PeerJ. 2018;6:e5716.

[56] Vijayakumar S, Menakha M. Pharmaceutical applications of cyanobacteria-A review. J Acute Med [Internet]. 2015;5(1):15-23. Available from: http://dx.doi.org/10.1016/j.jacme.2015.02.004

[57] Abed RMM, Dobretsov S, Sudesh K. Applications of cyanobacteria in biotechnology. J Appl Microbiol. 2009;106(1):1-12.

[58] Pagels F, Guedes AC, Amaro HM, Kijjoa A, Vasconcelos V. Phycobili proteins from cyanobacteria: Chemistry and biotechnological applications. Biotechnol Adv [Internet]. 2019;37(3):422-43. Available from: https://doi.org/10.1016/j.biotechadv.2019.02.010

[59] Telford WG, Moss MW, Morseman JP, Allnutt FCT. Cyanobacterial stabilized phycobilisomes as fluorochromes for extracellular antigen detection by flow cytometry. J Immunol Methods. 2001;254(1-2):13-30.

[60] Earthrise. What is Linablue [Internet]. [cited 2021 Jan 11]. Available from: https://www.earthrise.com/what-is-linablue

[61] Algae World News. The first algae-based ink is derived from spirulina instead of petrochemicals [Internet].

2019 [cited 2020 Dec 26]. Available from: https://news.algaeworld.org/2019/12/the-first-algae-based-ink-is-derived-from-spirulina-instead-of-petrochemicals/

[62] LGSonic. Ultrasonic Algae Control - Treat Large Water Surfaces - LG Sonic [Internet]. Available from: https://www.lgsonic.com/ultrasonic-algae-control/

[63] Cyano Biotech GmbH [Internet]. CyanoBiotech. Available from: http://www.cyano-biotech.com/content/home/index.php

[64] Photanol. Photanol [Internet]. Available from: https://photanol.com/

[65] Biotechnologia A. Algae biotechnologia [Internet]. Available from: http://www.algae.com.br/site/pt/

[66] Algenuity. Making algae work [Internet]. Available from: https://www.algenuity.com/

[67] LivingInkTech — About Living Ink [Internet]. Available from: https://livingink.co/about

[68] Spira. Spira Inc [Internet]. Available from: https://www.spirainc.com/

Chapter 3

Potential of Cyanobacteria in Wound Healing

Laxmi Parwani, Mansi Shrivastava and Jaspreet Singh

Abstract

The wound care market is rapidly expanding due to the development of innumerable dressings that exhibit specific healing requirements for different wound types. The use of biomaterials as suitable wound dressing material is highly advantageous due to their biocompatibility, biodegradability, and non-toxicity. Cyanobacteria have been widely explored for their potential applications in wound healing, as they are the rich source of bioactive compounds with antibacterial, antitumor, antiviral, antioxidant, and antifungal activities. In recent years this group of organisms has been widely studied due to their immense potential in biomedical applications. Although their different bioactivities can support wound healing in different ways, very few forms have proven utility as a wound-healing agent. This chapter gives an insight into the potential of cyanobacteria in wound healing. Different bioactive compounds present in variable forms of cyanobacteria and their associated activities were reported to support tissue regeneration and wound healing acceleration. As the demand for cost-effective, bioactive wound care products is ever increasing, these organisms have immense potential to be utilized for the development of bioactive wound dressings. Hence, various bioactive compounds of cyanobacteria, their associated activities, and roles in wound healing have been briefly reviewed in this chapter.

Keywords: Cyanobacteria, wound healing, bioactive compounds, hemostatic, antioxidant activities

1. Introduction

Wound healing involves various interactions between cellular, molecular, biochemical, and physiological activities, making the process very complex, dynamic, and precisely programmed. Wound healing involves four phases: hemostasis, inflammation, proliferation, and remodeling to restore the structural, functional, and physiological integrities of injured tissues [1]. This process results in the regeneration and replacement of injured tissue at the wound site [2]. Several nutritional factors are required for proper cellular differentiation, immune functioning, and collagen formation [3]. Any interruption, aberrance, or prolongation in the healing process would extend the tissue damage and thus prolong the repair process, contributing to chronic wound healing [4]. In recent years there has been accelerating demand for various wound dressings, each with specific characteristics, considering the distinctive healing requirements of various wound types. Various synthetic and natural products have been widely explored for their efficiency and accelerating wound healing abilities to accomplish the need for suitable dressing for a particular wound type.

The nonabsorbent and non-biodegradable nature of synthetic products makes these products unsuitable for healing purposes [5]. Medicinal plants are widely being explored for their utility in wound healing, and the ancient knowledge of medicinally essential plants, their increased popularity and utility further raised interest in exploring new natural products useful for the healing process. Various studies confirmed the anti-inflammatory, pro-collagen synthesis, antioxidant and antibacterial activities of natural products from plants and microbes, potentially beneficial for healing purposes. Biocompatibility and the presence of various bioactive phytochemicals in natural products efficiently promote the healing process and make them economically suitable for designing and fabricating dressings [6]. To satisfy the demand for new natural therapeutic agents for wound healing and to decrease the average costs involved in their development, researchers are screening organisms from overlooked microbial sources having the potential to be used as effective wound healing agents such as proteobacteria, bacterioidetes, and cyanobacteria.

Cyanobacteria are ubiquitous, oxygenic photosynthetic bacteria having diverse nature and can be found in various forms like unicellular and filamentous, marine and freshwater, free-living symbiotic, edible, and poisonous [7, 8]. This group of organisms has immense potential to produce many primary and secondary metabolites thus are known to perform potent biological activities. Produced secondary metabolites are low molecular weight, natural organic compounds, essential for average growth and development of these organisms. These metabolites have a wide range of applications in the field of medicines, industries, and biotechnology. These metabolites are a rich source of bioactive compounds, and cyanobacteria are the most promising organisms to produce them. In the last few decades, cyanobacteria have gained lots of attention for their medicinal values and wound healing properties. They are the choice of organisms due to their easy availability, fast regeneration, and huge diversity which have further expanded interest of researchers in their values as medicine and functional foods [9]. Thus, their potential as a good source of new therapeutic lead compounds has been realized during the last two decades. Cyanobacterial secondary metabolites show different medicinally essential activities such as antitumor, antibacterial, antifungal, antiviral, anti-inflammatory, immunomodulatory effects, and protease inhibition [10]. These biological activities are helpful to promote wound healing in different ways. Despite their potent biological activities, very few cyanobacterial forms are known to be useful in wound healing acceleration. Thus, this chapter presents an overview of bioactive compounds of cyanobacteria responsible for their various biological activities that promote the wound healing process by affecting different phases and factors of wound healing.

2. Important wound healing properties of cyanobacteria

2.1 Antibacterial activity

Antibacterial compounds are the molecules that kill or inhibit the growth of bacteria by affecting their physiological processes. Increased temperature due to inflammation caused by immune response and humidity caused by accumulation of fluid at the wound surface, make the wounds more vulnerable for bacterial infections. Once the wounded site came in contact with bacteria, they can penetrate underlying tissues, leading to life-threatening infections. To deal with such pathological conditions, effective wound management practices are essentially required. To improve the healing process and to reduce the wound bacterial colonization as well as infection at the wound site, the use of systematic antibiotics and the application of antimicrobial-loaded wound dressings are viable options to overcome the issue of infection and associated delay in healing. Considering the fact that usage of antibiotics has led to the increasing emergence of multidrug-resistant (MDR) strains of bacteria which negatively affects the healing process and worsens this issue [11]. Thus, there is a need to develop/discover an effective wound management material which can efficiently cure wound in such conditions. Several cyanobacteria are known to produce intracellular and extracellular bioactive compounds that possess antibacterial activities [12]. These compounds are effective against various bacterial strains like *Staphylococcus epidermidis*, *Bacillus cereus*, *Escherichia coli, Staphylococcus aureus, Salmonella typhi, Staphylococcus albus, Micrococcus flavus, Bacillus subtilis, Pseudomonas aeruginosa, Serratia marcescens, Mycobacterium smegmatis, Streptococcus pyogenes*, and some other Gram-positive and Gram-negative bacteria [13–19]. However, the production of antibacterial agents depends upon the composition of the culture medium, incubation period, pH, temperature, and light intensity during the growth of the selected cyanobacteria [20]. **Table 1** shows the bioactive compounds produced by different cyanobacterial strains, potentially useful for their antibacterial activity.

2.2 Antifungal activity

Devastating chronic wound infection is a significant reason for trauma worldwide, which causes serious public health problems. Majorly bacteria are responsible for infections in wounds, but very few recent studies analyzed both fungal and bacterial communities in the microbiome of chronic wounds, suggesting the role of fungi as underappreciated agents that leads to complications at the wound site [33].

Cyanobacterial Strain	Bioactive Compounds	Reference
Nostoc commune	Noscomin,Comnostins	[13, 21]
Nostoc sp.	Comnostin, Muscoride A, Dodecahydrophenanthrene, 4-methylchrysazin, Norharmane and 4-hydroxy-7-methylindan-1,carbamidocyclophane, Nostocarboline	[13, 16, 22]
Nostoc muscorum	Muscoride A	[23]
Lyngbya majuscula	Tanikolide	[24]
Microcoleus lacustris	Abietane	[25]
Oscillatoria redekei	Coriolic acid	[14]
Lyngbya sp.	Lyngbyazothrin, pahayokolide A	[26, 27]
Scytonema ocellatum, Tolypothrix conglutinate	Tolytoxin	[28]
Lyngbya majuscula	Pitipeptolides, malyngolid	[29, 30]
Scytonema sp.	Scytonemin	[31]
Nostocspongiaeforme	Tenuecyclamides	[31]
Micrococcus lacustris	Norbietane	[32]
Fischerella sp.	hapalindole T, ambiguine-I isonitrile	[14, 18]
Anabaena	Exopolysaccharide	[32]
Tolypothrix tenuis	Exopolysaccharide	[32]

Table 1.
Bioactive compounds of cyanobacterial strains possess antibacterial activity.

Hard to heal wounds like diabetic foot ulcers (DFUs) are majorly infected by members of the genus *Candid*a [34, 35]. Prolong infection along with delayed healing makes these wound types chronic and challenging to manage. Over 75% *Candid*a species were isolated by DFUs. In allergic rhinitis, respiratory disease, diabetic foot ulcers, other allergic fungi such as *Cladosporium* spp., *Aspergillus* spp., *Penicillium* spp., *Alternaria* spp., *Pleospora* spp., and *Fusarium* spp. are reported as critical infection-causing agents. Pathogenic and opportunistic fungi such as *Candida* spp., *Trichosporon asahii*, and *Rhodotorula* spp. are reported to associate with stalled open wounds or wounds resulting in an amputation [33]. Antifungal drugs to cure infected wounds are limited due to their high cost and side effects. So there is an emerging need to discover a new natural fungicidal agent. Many cyanobacterial compounds have been reported to possess inhibitory effects against different pathogenic strains of fungi [36]. They are an excellent source of various bioactive compounds that belong to several different chemical classes like peptides, polyketides, and alkaloids [37]. Bioactive antifungal compounds like nostodione, nostocyclamide, carazostatin, scytophycins, tolytoxin, phytoalexin, fisherellin, hapalindole are isolated from different strains of cyanobacteria [38]. Organic extracts of several cyanobacteria like *Oscillatoria latevirens, Chrococcus minor, Phormidium corium, Lyngbya martensiana, Lyngbya aestuarii, Aphanothece bullosacrude* and *Microcystis aeruginosa* have potential antifungal activity against *Candida albicans* and *Aspergillus flavus* [19]. **Table 2** shows various bioactive compounds of cyanobacteria possessing antifungal activity.

2.3 Antioxidant activity

Antioxidants are substances that prevent oxidation. They balance oxidative stress by eliminating free radicals and allowing the regeneration of tissue by repairing the cells [50]. Respiratory burst generates oxidants during wound healing; production of these oxidants supports acceleration in healing. These produced oxidants act as a messenger and thus promote healing. However, a delicate balance between oxidants and antioxidants required to control the progress of the wound. For normal wounds, low physiology levels of reactive oxygen species and oxidative stress are required at wound sites. Whereas oxidative stress and impaired wound healing led by their overexposure. To improve the level of oxidative stress, increased level of antioxidants is expected, which further help in healing acceleration [51]. Antioxidants also preserve

Cyanobacterial Strain	Bioactive Compounds	Reference
Scytonema hofmanni	Cyanobacterin	[39]
Calothrix fusca	Calophycin	[40]
Hapalosiphon fontinalis	Fontonamide, hapalindole	[38, 41]
Tolypothrix tenuis	Toyocamycin, Tubercidin	[42, 43]
Lyngbya majuscula	Majusculamide C, Hectochlorin, Tanikolide	[24, 44, 45]
Hyella caespitosa	Carazostatin	[46]
Nostoc sp.	Nostocyclamide	[47]
Nostoc commune	Nostofungicidine, Nostodione	[48, 49]
Scytonema sp.	tolytoxin, phytoalexin, scytophycins	[38]
Fischerella muscicola	fisherellin	[38]

Table 2.
Antifungal activity of bioactive compounds isolated from different cyanobacterial strains.

and stimulate the function of immune cells against homeostatic disorders. Therefore, their increased levels can improve the immune response and accelerate wound healing [52]. Accumulation of low molecular weight iron at the wound site also increases inflammation and microbial invasion of the wound [53], suggesting the requirement of a chelating agent to achieve proper healing. Production of various, chemically diverse groups of secondary metabolites from cyanobacteria established their industrial significance and made them an excellent source of antioxidants that facilitate the formation of the body's defense mechanism against free radical induced damages to cells. Their antioxidant and metal chelating ability are reportedly due to phytonutrients and pigments present in them [54]. Their cell-free extracts possess free radical scavenging property, metal chelating activity, and deoxyribose protection [55]. The antioxidant properties of the cyanobacterial cell extracts are imparted by the total phenol and total flavonoid content present in the extracts. The free radical scavenging, metal chelating, and antioxidative damage protecting properties of cyanobacterial cell extracts are presumably linked with varied quantities of polyphenolics, gallic, chlorogenic, caffeic, vanillic, and ferulic acids, flavonoids, quercetin, and kaempferol present in them. Exopolysaccharides of cyanobacteria also exhibit good antioxidant and anticoagulant activities. They can also induce oxidants and antioxidant enzymes and are known as immunostimulators [56]. Exopolymers of three strains of *Anabaena* and *Tolypothrix tenuis* exhibited antioxidant activities against O_2·, H_2O_2, OH·, and NO·. These polymers also possess $Fe2^+$ chelating activity, helps in preventing the invasion of pathogenic organisms at the wound site. A strong correlation of sulfate content against superoxide and nitric oxide radicals scavenging activity of exopolymers was found, whereas H_2O_2 scavenging ability and reducing power were contributed by phenols present in them. The overall reducing power and superoxide control strongly related to their iron chelation ability [57]. The highest phenolic, flavanoid, and phycobiliprotein content in *Lyngbya* sp. possessed the highest 2,2-diphenyl-1-picrylhydrazyl (DPPH) activity. The antioxidant potential of *Spirulina* sp. was presumable related to the presence of alcoholic and phenolic OH groups in the cellular structure of the organism [54]. Phenolic content in methanolic extracts of *Anabaena* sp., *Nostoc* sp., *Nostoc commune, Nodularia spumigena, Leptolyngbya protospira, Phormidiochaete* sp. and *Arthrospira* platensis reported higher antioxidant activity than their aqueous extracts [58, 59]. Similarly, ethanolic extracts of *Phormidium fragile, Lyngbya limnetica, Scytonema bohnerii* and *Calothrix fusca* possess antioxidant ability [60]. Cyanobacterial pigments also possess antioxidant activities. In *Lyngbya* and *Phormidium* sp. peroxyl and hydroxyl radicals, scavenging ability was reported to be linked with the presence of covalently linked tetrapyrrole chromophore with phycocyanobilin [61]. Phycocyanin, a type of phycobiliprotein isolated from *Spirulina platensis* and *Geitlerinema* sp., reported to have antioxidant and anti-inflammatory properties and ability to scavenge peroxyl, hydroxyl, and superoxide radicals [62]. Phytonutrients, pigments, and polysaccharides of cyanobacteria exhibit good antioxidant, anticoagulant, and metal chelating ability [54, 56]. Polysaccharidic exopolymers of three strains of *Anabaena* and *Tolypothrix tenuis* possessed antiradical and Fe^{2+} chelating activity. All these exopolymers exhibited antioxidant activities against $O_2^{\bullet}$, H_2O_2, $OH^{\bullet}$, and $NO^{\bullet}$, thus helps in the rapid healing of the wound [57].

2.4 Immunomodulatory and anti-inflammatory effects

Inflammation is a local, protective and physiological response to microbial invasion or injury. The magnitude of the inflammatory response is crucial: insufficient responses result in immunodeficiency, whereas excessive responses cause morbidity and mortality. Therefore, homeostasis and health are restored when inflammation is limited by anti-inflammatory responses [63]. Natural compounds have gained lots of attention in treating various types of inflammations to reduce the reaction of the immune system against pathogens, toxic compounds, and damaged cells. The immune system actively participates in homeostasis, re-establishment, following tissue injury via multiple mechanisms, and plays a critical role throughout the wound healing process. Immune system control to promote tissue repair and regeneration is an attractive approach when designing regenerative strategies. Now a day, the multifunctional immunomodulatory properties of cyanobacteria are gaining much attention in the field of medicine. Cyanobacteria produce various metabolites with different chemical structures, including small molecules of peptides and proteins, polysaccharides, fatty acids, and their derivatives, possessing anti-inflammatory activities [64]. Different cyanobacterial components control the release of certain cytokines from human monocytes and macrophages. Depending on the wound microenvironment, they also can reduce or activate the production of reactive oxygen species from neutrophils [65]. Immunomodulatory effects of cyanobacteria highly induce activation of both types of immune cells which could promote wound debridement, accelerate re-epithelization, and wound closure [66]. Edible cyanobacterial forms are known for their immune-boosting abilities, and many studies have proven their immunomodulatory and anti-inflammatory effects. *Spirulina* is a widely consumed cyanobacterial strain because of its extraordinary neutraceutical and pharmaceutical values. Dietary effects of *Spirulina platensis* showed increased phagocytic and increased antigen production and increased natural killer cell-mediated antitumor activity in the test animals under study [67]. Daily consumption of *Spirulina* stimulates and promotes the immune system by increasing the phagocytic activity of macrophages, induces antibodies and cytokines production, increases accumulation of natural killer cells into tissues, and activates T and B cells [68]. Daily consumption of *Spirulina* for 16 consecutive weeks reported a significant rise in plasma IL- 2 concentration and a significant reduction in IL-6 concentration in humans [69]. Nonadecane and 9-Eicosyne in *Spirulina* were supposed to be responsible for the anti-inflammatory effects [70]. A high-molecular-weight polysaccharide extracted from the cyanobacterium *Spirulina* is known as Immulina®, a potent activator of innate immune cells and exerts inhibitory effects in the induced allergic inflammatory response [71, 72]. It also exhibits anti-inflammatory properties and can inhibit the release of histamine from mast cells [47]. Gama and alpha-linolenic acid are the fatty substances extracted from *Spirulina* and *Aphanizomenon flos-aquae*, respectively, which inhibit the formation of inflammatory mediators [48]. Similarly, *Nostoc* is an edible, largely consumed cyanobacterial form after *Spirulina.* Polysaccharide-rich extracts of *Nostoc commune* could be potentially used for macrophage activation and consequently inhibit the leukemic cell growth and induce monocytes/macrophages when used to treat human monoblastoid leukemia U937 cells [73]. Different immunomodulating activities of *Microcystis aeruginosa, Synechocystis aquatilis, Oscillatoria redekei, Anabaena flos-aque, Aphanizomenon flos-aquae, Oscillatoria rubescens, Oscillatoria tenuis* have also been reported [74] their different cell extracts variably promote the proliferation of lymphocytes. On the other hand, some species of cyanobacteria that contain cytotoxic metabolites (cyanotoxins) can cause immunotoxicity and immunosuppression. The property of immunotoxicity is well reported in cyanobacterial bloom extracts containing microcystin [75]. Treatment with microcystin showed inhibition of lipopolysaccharide-induced lymphoproliferation and decreased numbers of antibody-forming cells; this results in immunosuppression in mice that were immunized using T-dependent antigen sheep red blood cells. A form of cyanobacterial toxin, cylindrospermopsin, is an important water pollutant having broad biological activity. Cylindrospermopsin promotes significant production of pro-inflammatory mediator tumor necrosis factor α and reactive oxygen species when introduced in macrophages cells [76]. Effmert in 1991 reported aqueous extracts of different cyanobacterial species *Microcystis aeruginosa, Synechocystis aquatilis, Oscillatoria redekei, Anabaena flos-aque, Aphanizomenon flos-aquae, Oscillatoria rubescens, Oscillatoria tenuis* possess strong immunomodulating activities [74]. Aqueous extract of the marine cyanobacterium, *Trichodesmium erythraeum*, ethyl acetate extract of non-edible blue-green algae *Geitlerinema splendidum* exhibit anti-inflammatory activity [77, 78]. In chronic wounds where a coordinated wound healing cascade fails to progress beyond the inflammatory phase, the anti-inflammatory activity of cyanobacteria helps in the wound healing by proliferation, matrix deposition, and ultimately, wound resolution [79]. During the last decade, several bioactive compounds have been isolated from cyanobacteria having anti-inflammatory properties that suggest their importance and potential in developing a huge market of wound healing aids [10]. Different classes of cyanobacterial bioactive compounds like phycobilins, phenols, polysaccharides, steroids, and terpenoids possess potential anti-inflammatory effects. Selected anti-inflammatory compounds of different cyanobacterial strains are listed in **Table 3**.

2.5 Hemostatic activity

For people throughout the world, traumatic injuries have been a challenge. Considering technological advancements made through age's trauma remains a leading cause of human morbidity and mortality [90]. Excess bleeding can cause delayed wound healing, hematoma formation, infection, dehiscence, and necrosis. Patients suffering from trauma and its consequent hemorrhage essentially require the establishment of hemostasis by topical wound dressings. Constriction of blood vessels, the activation of the coagulation cascade, and the formation of blood clots are essential

Chemical group of bioactive compound	Bioactive Compound	Cyanobacterial Strain	References
Amino acids and peptides	Aeruginosin	*Nostoc* sp.	[80]
	Porphyra	*Aphanizomenon flos-aquae*	[81]
	Shinorine	*Anabaena variabilis*	[82]
	Phycocyanin	*Spirulina* sp	[72]
	Ethyl tumonoate A	*Oscillatoria margaritifera*	[83]
	Cyanopeptolin	*Microcystis* spp	[84]
Polysaccharide	Sacran	*Aphanothece sacrum*	[85]
Lipids	Monogalactosyl diacylglycerol, Digalactosyl diacylglycerol, Sulphoquinovosyl diacylglycerol, Phosphatidyl glycerol	*Phormidium* sp.	[86]
Pigment	Scytonemin	*Scytonema*	[87]
Others	Coibacin A	*Oscillatoria* sp.	[88]
	Honaucins A–C	*Leptolyngbya crossbyana*	[89]

Table 3.
Bioactive compounds of cyanobacterial strains possessing anti-inflammatory properties.

and significant steps of hemostasis. Therefore, any effort made to accelerate any or all phases above can help achieve hemostasis [91]. The role of cyanobacteria and algae in hemostasis is significantly less known and identified. In the majority, their antithrombic activities useful to cure thrombosis-related diseases are reported. The antithrombic activity of Spirulan, a sulfated polysaccharide of *Arthrospira platensis* has been widely known. Spirulan is helpful to directly decrease the activity of thrombin and factor X activated, procoagulant proteins that can be used to prevent thrombus formation and partial lysis of thrombus [92]. Similarly, *Microcystis aeruginosa* and different cyanobacteria blooms produce probable fVIIa-soluble Tissue Factor (fVIIa-sTF) inhibitors [93]. Vitamin-K-dependent clotting factors like thrombin and fVIIa are associated with bleeding, related complications, and disorders. These factors can induce excessive bleeding when treated with vitamin-K antagonists. *M. aeruginosa* produces bioactive compound Aeruginosin, which shows positive fVIIa-sTF inhibitory activity. Daily consumption of aqueous extract of *Spirulina platensis*, containing a high dose of phycocyanin was found safe for when tried on the human to test its anticoagulant activity and platelet activation ability. The studied extract is also suitable for providing rapid and robust relief in chronic pain. The extract also improved the liver function and metabolism by reducing the activity of aspartate transaminase and alanine transaminase enzymes [94]. C-phycocyanins extracted from *Spirulina platensis* possess significant anticoagulation, antioxidant, and prevention of DNA damage activities [95]. Exopolysaccharides of *A. platensis* possess antiatherogenic and anti-thrombogenic activities [96]. Although the significant roles of cyanobacterial and algal polysaccharides as antioxidants, antiviral, antitumoral, and anticoagulant have been well-documented and reviewed [10] the single report was found to date mentioning the role of cyanobacterial exopolymers (EPs) isolated from four desert cyanobacteria (*Tolypothrix tenuis* and three species of *Anabaena*) in hemostasis. These exopolymers were potentially beneficial to reduce activated partial thromboplastin time (APTT, a measure of how long blood takes to clot) and prothrombin time (PT, a measure of how long blood plasma takes to clot) by 16–41% and 12–65% respectively. The gravimetric method of thrombogenicity assessment showed that the blood clot formed by the cyanobacterial EPs was heavier vis a vis glass (positive control) and thus was thrombogenic. Similar studies can open the treasure of hidden potential of bioactive compounds of cyanobacteria in hemostasis which needs to be explored more.

2.6 Wound healing effect of cyanobacteria

Specific healing requirements of a wound widen the scope of identifying more natural, economic, and effective wound healing agents, which gave extraordinary rise to the development of many synthetic and natural products useful as suitable wound healing dressing materials. Various natural products obtained by plants are widely known for their medicinal properties, facilitating wound healing. Bioactive secondary metabolites like alkaloids, essential oils, flavonoids, tannins, terpenoids, saponins, and phenolics are present in plant-based natural products. These compounds possess various activities like anti-inflammatory, antioxidant, antibacterial, procollagen synthesis, etc., and efficacy to modulate one or more phases of the wound healing process which further help in accelerated tissue regeneration during healing [97]. Similarly, cyanobacterial bioactive compounds possess important medicinal properties like immunostimulating, antiviral, antioxidant, antibacterial, antifungal, antialgal, and anticancerous activities [38, 98] but very few are known for their wound healing potential. The blue-green microalgae *Spirulina* has been widely studied for its wound healing potential and possess antioxidant, antimutagenic, antiviral, anticancer, anti-allergic, immune-enhancing, hepatoprotective, blood vessel relaxing, and blood lipid-lowering activities [99, 100]. The C-phycocyanin pigment of *Spirulina* promoted proliferation, regeneration, and migration of cells tested on cultured human keratinocytes. A similar set of experiments on Sprague–Dawley male rats showed re-epithelization, neovascularization, presence of inflammatory cells, granulation of tissue amount, and maturation during wound healing [101]. Aqueous extract of *Spirulina platensis* showed proliferation, migration, and enhanced closure rate of wound area within 24 hours when tested on human dermal fibroblast cells (HDF) [102]. Compounds like cinnamic acid, narigenin, kaempferol, temsirolimus, phosphatidylserine, and isomeric derivatives were presumably involved in the accelerated wound healing activity of the studied aqueous extract of *Spirulina platensis*. Antibacterial and antioxidant activities of *Spirulina* are useful to improve the immunity of skin cells, and thus burn cream of ethanolic extract of *Spirulina* along with gold nanoparticles has been formulated. To reduce the pain by numbing the injured area, Lidocaine has been added to the cream as an anesthetic, sodium dodecyl sulfate, and glycerine used for skin moisturization [103]. Polylactic acid (PLA), polyethylene oxide, and PHB (polyhydroxybutyrate) extracted from *Arthrospira* LEB 18 strain used to prepare biodegradable, biocompatible nanofibrous scaffolds. Properties of prepared scaffolds like increased conductivity, higher mechanical durability with enhanced elasticity, tensile strength, and breaking elongation supported nutrient, growth factors, and metabolism byproducts distribution at the wound site [104]. Scaffolds of poly-D, L-lactic acid (PDLLA) associated with LEB 18 biomass showed better adherence to the wound, increased cell viability, and more moldable when compared with classic PDDLA [105]. Natural wound healing processes can be promoted by such biomatrices produced from *Arthrospira*, and also they can minimize the risk of infection [104]. PCL nanofiber loaded with *Spirulina* extract was fabricated as a cutaneous wound dressing material. The designed dressing showed enhanced wound regeneration ability by modulating intra- and extracellular ROS with enhancing antioxidant mechanism and increased fibroblast viability under oxidative stress [106]. *Synechococcus elongatusa*, a naturally occurring photoautotrophic cyanobacterium promoted angiogenesis and burn wound repair in mice. The promotion of interleukin-6 expression and secretion of extracellular vesicles probably induced pro-angiogenic and wound healing effects in the studied animals [107]. In diverse marine, freshwater, and terrestrial organisms, ultraviolet-absorbing mycosporine-like amino acids (MAAs) are found as secondary metabolites. *Nostoc flagelliforme* and *Nostoc commune*, are widely known for their edible values and reported as natural resources of MAAs. These secondary metabolites possess several beneficial effects and are useful as sun-screening cosmetics, antioxidants, and pharmaceuticals [108]. In wound healing, MAAs modulate skin fibroblast proliferation and activate adhesion kinases, extracellular signal-regulated kinases that promote acceleration in wound healing. Extensive researches are conducting on MAAs to identify their potential as a new wound healing agent to be explored as a novel biomaterial for wound healing therapies [109]. Cyanobacterial glycans are known as hydrogels that imbibe large amounts of biofluids [110]. This property of glycans extremely useful for targeted drug delivery in chronic wound cases [111], where they have been exploited for delivery of low molecular weight drugs and macro-molecular payloads (hormones, peptide, and protein drugs) [112]. Cyanobacteria have been widely explored and studied for their immense potential in various biomedical applications, but very few are reported for their accelerated wound healing abilities. Most of the studies have been conducted on *Spirulina* which is known for its extraordinary neutraceutical and pharmaceutical values. As these organisms are phototrophic and have minimal growth requirements, they can be easily regenerated, and also their huge diversity makes them suitable candidates for research, having hidden potential for wound healing. Further, they can be explored to develop bioactive and cost-effective wound dressing materials, suitable for all sections of society.

3. Conclusions

The ever-increasing demand for effective bioactive dressings suitable for different types of wounds is rapidly expanding the wound care market at diverse levels. Cyanobacteria are known for their numerous biomedical applications; thus, recently, they are a widely explored group of organisms. The cyanobacterial bioactive compounds possess antiviral, antifungal, antibacterial, antitumor, anti-inflammatory, and antioxidant activities suitable to accelerate wound healing even in chronic conditions by controlling different phases and factors of the wound. Cyanobacteria are known for the abundant availability of versatile bioactive compounds and their associated properties, but unfortunately, very few forms of cyanobacteria are known for their role in wound healing. Their easy cultivation, colossal diversity, and different biological activities can make them suitable candidates for research. Further, they can be explored more in the field of biomaterials for designing and fabrication of low-cost, biocompatible, and biodegradable wound dressings.

Acknowledgements

The authors are indebted to the Department of Science and Technology (DST), Govt. of India New Delhi, for providing opportunity and financial support (SP/YO/2019/1502) for interdisciplinary research work. The authors are also grateful to Professor Ashish Bhatnagar and Professor Monica Bhatnagar, Department of Microbiology, M.D.S. University, Ajmer, India for their support in the subject study.

Conflict of interest

The authors declare no conflict of interest.

Author details

Laxmi Parwani[1]*, Mansi Shrivastava[1] and Jaspreet Singh[2]

1 Department of Bioscience and Biotechnology, Banasthali Vidyapith, P.O. Banasthali, Rajasthan, India

2 School of Life Sciences, SIILAS Campus, Jaipur National University, Jaipur, Rajasthan, India

*Address all correspondence to: laxmi_parwani@rediffmail.com

References

[1] Judith R, Nithya M, Rose C, Mandal AB. Application of a PDGF containing novel gel for cutaneous wound healing. Life Sciences. 2010;87: 1-8. DOI: 10.1016/j.lfs.2010.05.003

[2] Velnar T, Bailey T, Smrkolj V. The wound healing process: an overview of the cellular and molecular mechanisms. Journal of International Medical Research. 2009;37(5):1528-1542. DOI: 10.1177/147323000903700531

[3] Mackay D, Miller AL. Nutritional support for wound healing. Alternative Medicine Review. 2003;8:359-377. DOI: 10.1002/jor.23656

[4] Kurahashi T, Fujii J. Roles of antioxidative enzymes in wound healing. Journal of Developmental Biology. 2015;3(2):57-70. DOI: https://doi.org/10.3390/jdb3020057

[5] Mir M, Ali MN, Barakullah A, Gulzar A, Arshad M, Fatima S, Asad M. Synthetic polymeric biomaterials for wound healing: a review. Progress in biomaterials. 2018;7(1):1-21. DOI: 10.1007/s40204-018-0083-4

[6] Thakur R, Jain N, Pathak R, Sandhu SS. Practices in wound healing studies of plants. Evidence-based complementary and alternative medicine. 2011;2011:1-17. DOI: https://doi.org/10.1155/2011/438056

[7] Taton A, Grubisic S, Balthasart P, Hodgson DA, Laybourn-Parry J, Wilmotte A. Biogeographical distribution and ecological ranges of benthic cyanobacteria in East Antarctic lakes. FEMS microbiology ecology. 2006;57(2):272-289. DOI: https://doi.org/10.1111/j.1574-6941.2006.00110.x

[8] Dorador C, Vila I, Imhoff JF, Witzel KP. Cyanobacterial diversity in Salar de Huasco, a high altitude saline wetland in northern Chile: an example of geographical dispersion? FEMS microbiology ecology. 2008;64(3):419-432. DOI: 10.1111/j.1574-6941.2008.00483.x

[9] Barrow C, Shahidi F. Marine nutraceuticals and functional foods.1st ed. CRC Press; 2007. 512 p. DOI: https://doi.org/10.1201/9781420015812

[10] Singh R, Parihar P, Singh M, Bajguz A, Kumar J, Singh S, Singh VP, Prasad SM. Uncovering potential applications of cyanobacteria and algal metabolites in biology, agriculture and medicine: current status and future prospects. Frontiers in microbiology. 2017;8:515. DOI: https://doi.org/10.3389/fmicb.2017.00515

[11] Li B, Webster TJ. Bacteria antibiotic resistance: New challenges and opportunities for implant-associated orthopedic infections. Journal of Orthopaedic Research®. 2018;36(1):22-32. DOI: 10.1002/jor.23656

[12] Schlegel I, Doan NT, de Chazal N, Smith GD. Antibiotic activity of new cyanobacterial isolates from Australia and Asia against green algae and cyanobacteria. Journal of Applied Phycology. 1998;10(5):471-479. DOI: https://doi.org/10.1023/A:1008042619686

[13] Ostensvik O, Skulberg OM, Underdal B, Hormazabal V. Antibacterial properties of extracts from selected planktonic freshwater cyanobacteria-A comparative study of bacterial bioassays. Journal of applied microbiology.1998;84(6):1117-1124. DOI: 10.1046/j.1365-2672.1998.00449.x

[14] Mundt S, Kreitlow S, Jansen R. Fatty acids with antibacterial activity from the cyanobacterium *Oscillatoria redekei* HUB 051. Journal of Applied Phycology. 2003;15(2):263-267. DOI: https://doi.org/10.1023/A:1023889813697

[15] Bui HT, Jansen R, Pham HT, Mundt S. Carbamidocyclophanes A-E, chlorinated paracyclophanes with cytotoxic and antibiotic activity from the Vietnamese cyanobacterium *Nostoc* sp. Journal of natural products. 2007;70(4):499-503. DOI: 10.1021/np060324m

[16] Nowruzi B, Haghighat S, Fahimi H, Mohammadi E. *Nostoc* cyanobacteria species: a new and rich source of novel bioactive compounds with pharmaceutical potential. Journal of Pharmaceutical Health Services Research. 2018;9(1):5-12. DOI: https://doi.org/10.1111/jphs.12202

[17] El-Sheekh MM, Osman ME, Dyab MA, Amer MS. Production and characterization of antimicrobial active substance from the cyanobacterium *Nostoc muscorum*. Environmental toxicology and pharmacology. 2006;21(1):42-50. DOI: 10.1016/j.etap.2005.06.006

[18] Raveh A, Carmeli S. Antimicrobial ambiguines from the cyanobacterium *Fischerella* sp. collected in Israel. Journal of natural products. 2007;70(2):196-201. DOI: 10.1021/np060495r

[19] Ananya AK, Ahmad IZ. Cyanobacteria"the blue green algae" and its novel applications: A brief review. International Journal of Innovation and Applied Studies. 2014;7(1):251-261

[20] Noaman NH, Fattah A, Khaleafa M, Zaky SH. Factors affecting antimicrobial activity of *Synechococcus leopoliensis*. Microbiological Research. 2004;159(4):395-402. DOI: 10.1016/j.micres.2004.09.001

[21] Jaki B, Orjala J, Heilmann J, Linden A, Vogler B, Sticher O. Novel extracellular diterpenoids with biological activity from the cyanobacterium *Nostoc commune*. Journal of natural products. 2000;63(3):339-343. DOI: 10.1021/np9903090

[22] Becher PG, Keller S, Jung G, Süssmuth RD, Jüttner F. Insecticidal activity of 12-epi-hapalindole J isonitrile. Phytochemistry. 2007;68(19):2493-2497. DOI: 10.1016/j.phytochem.2007.06.024

[23] Nagatsu A, Kajitani H, Sakakibara J. Muscoride A: A new oxazole peptide alkaloid from freshwater cyanobacterium *Nostoc muscorum*. Tetrahedron Letters. 1995;36(23):4097-4100. DOI: https://doi.org/10.1016/0040-4039(95)00724-Q

[24] Pandey VD. Cyanobacterial natural products as antimicrobial agents. International Journal of Current Microbiology and Applied Sciences. 2015;4(1):310-317.

[25] Thajuddin N, Subramanian G. Cyanobacterial biodiversity and potential applications in biotechnology. Current science. 2005;89(1):47-57. DOI: https://www.jstor.org/stable/24110431

[26] Asthana RK, Srivastava A, Singh AP, Singh SP, Nath G, Srivastava R, Srivastava BS. Identification of an antimicrobial entity from the cyanobacterium *Fischerella* sp. isolated from bark of *Azadirachta indica* (Neem) tree. Journal of applied phycology. 2006;18(1):33-39. DOI: https://doi.org/10.1007/s10811-005-9011-9

[27] Zainuddin EN, Jansen R, Nimtz M, Wray V, Preisitsch M, Lalk M, Mundt S. Lyngbyazothrins A– D, Antimicrobial cyclic undecapeptides from the cultured cyanobacterium *Lyngbya* sp. Journal of natural products. 2009;72(8):1373-1378. DOI: 10.1021/np8007792

[28] Moore RE, Patterson GM, Mynderse JS, Barchi J, Norton TR, Furusawa E, Furusawa S. Toxins from cyanophytes belonging to the *Scytonemataceae*. Pure and Applied

Chemistry. 1986;58(2):263-271. DOI: 10.1351/pac198658020263

[29] Berry JP, Gantar M, Gawley RE, Wang M, Rein KS. Pharmacology and toxicology of pahayokolide A, a bioactive metabolite from a freshwater species of *Lyngbya* isolated from the Florida Everglades. Comparative Biochemistry and Physiology Part C: Toxicology & Pharmacology. 2004;139(4):231-238. DOI: 10.1016/j.cca.2004.11.005

[30] Helms GL, Moore RE, Niemczura WP, Patterson GM, Tomer KB, Gross ML. Scytonemin A, a novel calcium antagonist from a blue-green alga. The Journal of Organic Chemistry. 1988;53(6):1298-1307. DOI: https://doi.org/10.1021/jo00241a033

[31] Gutiérrez RM, Flores AM, Solís RV, Jimenez JC. Two new antibacterial norabietane diterpenoids from cyanobacteria, *Microcoleous lacustris*. Journal of natural medicines. 2008;62(3):328-331. DOI: 10.1007/s11418-008-0238-z

[32] Bhatnagar M, Parwani L, Sharma V, Ganguly J, Bhatnagar A. Exopolymers from *Tolypothrix tenuis* and three *Anabaena* sp. (Cyanobacteriaceae) as novel blood clotting agents for wound management. Carbohydrate polymers. 2014;99:692-699. DOI: 10.1016/j.carbpol.2013.09.005

[33] Kalan L, Grice EA. Fungi in the wound microbiome. Advances in wound care. 2018;7(7):247-255. DOI: 10.1089/wound.2017.0756

[34] Chellan G, Shivaprakash S, Ramaiyar SK, Varma AK, Varma N, Sukumaran MT, Vasukutty JR, Bal A, Kumar H. Spectrum and prevalence of fungi infecting deep tissues of lower-limb wounds in patients with type 2 diabetes. Journal of clinical microbiology. 2010;48(6):2097-2102. DOI: 10.1128/JCM.02035-09

[35] Dowd SE, Delton Hanson J, Rees E, Wolcott RD, Zischau AM, Sun Y, White J, Smith DM, Kennedy J, Jones CE. Survey of fungi and yeast in polymicrobial infections in chronic wounds. Journal of wound care. 2011;20(1):40-47. DOI: 10.12968/jowc.2011.20.1.40

[36] Pawar ST, Puranik PR. Screening of terrestrial and freshwater halotolerant cyanobacteria for antifungal activities. World Journal of Microbiology and Biotechnology. 2008;24(7):1019-1025. DOI: 10.1007/s11274-007-9565-6

[37] Shishido TK, Humisto A, Jokela J, Liu L, Wahlsten M, Tamrakar A, Fewer DP, Permi P, Andreote AP, Fiore MF, Sivonen K. Antifungal compounds from cyanobacteria. Marine drugs. 2015;13(4):2124-2140. DOI: https://doi.org/10.3390/md13042124

[38] Abed RM, Dobretsov S, Sudesh K. Applications of cyanobacteria in biotechnology. Journal of applied microbiology. 2009;106(1):1-2. DOI: 10.1111/j.1365-2672.2008.03918.x

[39] Mason CP, Edwards KR, Carlson RE, Pignatello J, Gleason FK, Wood JM. Isolation of chlorine-containing antibiotic from the freshwater cyanobacterium Scytonema hofmanni. Science. 1982;215(4531):400-402. DOI: 10.1126/science.6800032

[40] Moon SS, Chen JL, Moore RE, Patterson GM. Calophycin, a fungicidal cyclic decapeptide from the terrestrial blue-green alga *Calothrix fusca*. The Journal of Organic Chemistry. 1992;57(4):1097-1103. DOI: https://doi.org/10.1021/jo00030a013

[41] Moore RE, Yang XQ, Patterson GM. Fontonamide and anhydrohapaloxindole A, two new alkaloids from the blue-green alga *Hapalosiphon fontinalis*. The Journal of Organic Chemistry. 1987;52(17):3773-3777. DOI: https://doi.org/10.1021/jo00226a009

[42] Moore RE. Toxins, anticancer agents, and tumor promoters from marine prokaryotes. Pure and applied chemistry. 1982;54(10):1919-1934. DOI: https://doi.org/10.1351/pac198254101919

[43] Banker R, Carmeli S. Tenuecyclamides A– D, Cyclic Hexapeptides from the Cyanobacterium *Nostoc s pongiaeforme* var. t enue. Journal of natural products. 1998;61(10):1248-1251. DOI: 10.1021/np980138j

[44] Swain SS, Paidesetty SK, Padhy RN. Antibacterial, antifungal and antimycobacterial compounds from cyanobacteria. Biomedicine & Pharmacotherapy. 2017;90:760-776. DOI: 10.1016/j.biopha.2017.04.030

[45] Singh IP, Milligan KE, Gerwick WH. Tanikolide, a toxic and antifungal lactone from the marine cyanobacterium *Lyngbya m ajuscula*. Journal of natural products. 1999;62(9):1333-1335.DOI: https://doi.org/10.1021/np990162c

[46] Burja AM, Banaigs B, Abou-Mansour E, Burgessd JG, Wrighta PC. Marine cyanobacteria - A prolific source of natural products. Tetrahedron. 2001;57:9347-9377. DOI: https://doi.org/10.1016/S0040-4020(01)00931-0

[47] Moore RE, Patterson GM, Carmichael WW. New pharmaceuticals from cultured blue-green algae. Biomedical importance of marine organisms. 1988;13(1988):143-150. DOI: https://doi.org/10.1007/BF00003544

[48] Kajiyama SI, Kanzaki H, Kawazu K, Kobayashi A. Nostofungicidine, an antifungal lipopeptide from the field-grown terrestrial blue-green alga *Nostoc commune*. Tetrahedron letters. 1998;39(22):3737-3740. DOI: https://doi.org/10.1016/S0040-4039(98)00573-5

[49] Bhadury P, Wright PC. Exploitation of marine algae: biogenic compounds for potential antifouling applications. Planta. 2004;219(4):561-578. DOI: 10.1007/s00425-004-1307-5

[50] Houghton PJ, Hylands PJ, Mensah AY, Hensel A, Deters AM. In vitro tests and ethnopharmacological investigations: wound healing as an example. Journal of Ethnopharmaco logy. 2005;100(1-2):100-107. DOI: 10.1016/j.jep.2005.07.001

[51] Campa-Córdova AI, Hernández-Saavedra NY, Ascencio F. Superoxide dismutase as modulator of immune function in American white shrimp (*Litopenaeus vannamei*). Comparative biochemistry and physiology part C: Toxicology & pharmacology. 2002;133(4):557-565. DOI: 10.1016/s1532-0456(02)00125-4

[52] De la Fuente M, Victor VM. Anti-oxidants as modulators of immune function. Immunology and cell biology. 2000;78(1):49-54. DOI: 10.1046/j.1440-1711.2000.00884.x

[53] Weinberg ED. Iron withholding: a defense against infection and neoplasia. Physiological reviews. 1984;64(1):65-102. DOI: 10.1152/physrev.1984.64.1.65

[54] Hossain MF, Ratnayake RR, Meerajini K, Wasantha Kumara KL. Antioxidant properties in some selected cyanobacteria isolated from fresh water bodies of Sri Lanka. Food science & nutrition. 2016;4(5):753-758. DOI: 10.1002/fsn3.340

[55] Singh DP, Prabha R, Verma S, Meena KK, Yandigeri M. Antioxidant properties and polyphenolic content in terrestrial cyanobacteria. 3 Biotech. 2017;7(2):1-4.

[56] Costa LS, Fidelis GP, Cordeiro SL, Oliveira RM, Sabry DD, Câmara RB, Nobre LT, Costa MS, Almeida-Lima J, Farias EH, Leite EL. Biological activities

of sulfated polysaccharides from tropical seaweeds. Biomedicine & Pharmacotherapy. 2010;64(1):21-28. DOI: 10.1016/j.biopha.2009.03.005

[57] Parwani L, Bhatnagar M, Bhatnagar A, Sharma V. Antioxidant and iron-chelating activities of cyanobacterial exopolymers with potential for wound healing. Journal of applied phycology. 2014;26(3):1473-1482. DOI: 10.1007/s10811-013-0180-7

[58] Pant G, Kumar G, Karthik L, Prasuna RG, Rao KB. Antioxidant activity of methanolic extract of blue green algae *Anabaena* sp. (Nostocaceae). European Journal of Experimental Biology. 2011;1(1):156-162.

[59] Jerez-Martel I, García-Poza S, Rodríguez-Martel G, Rico M, Afonso-Olivares C, Gómez-Pinchetti JL. Phenolic profile and antioxidant activity of crude extracts from microalgae and cyanobacteria strains. Journal of Food Quality.2017;2017:1-8. DOI: https://doi.org/10.1155/2017/2924508

[60] Sharathchandra K, Rajashekhar M. Antioxidant activity in the four species of cyanobacteria isolated from a sulfur spring in the Western Ghats of Karnataka. International Journal of Pharma and Bio Sciences. 2013;4(1):275-285.

[61] Gerwick WH, Coates RC, Engene N, Gerwick L, Grindberg RV, Jones AC, Sorrels CM. Giant marine cyanobacteria produce exciting potential pharmaceuticals. Microbe-American Society for Microbiology. 2008;3(6):277-284. DOI: 10.1128/microbe.3.277.1

[62] Bermejo P, Piñero E, Villar ÁM. Iron-chelating ability and antioxidant properties of phycocyanin isolated from a protean extract of *Spirulina platensis*. Food chemistry. 2008;110(2):436-445. DOI: 10.1016/j.foodchem.2008.02.021

[63] Tracey KJ. The inflammatory reflex. Nature. 2002;420(6917):853-859. DOI: 10.1038/nature01321

[64] Tabarzad M, Atabaki V, Hosseinabadi T. Anti-inflammatory Activity of bioactive compounds from microalgae and cyanobacteria by focusing on the mechanisms of action. Molecular Biology Reports. 2020;47(8):6193-6205. DOI: 10.1007/s11033-020-05562-9

[65] Stewart I, Schluter PJ, Shaw GR. Cyanobacterial lipopolysaccharides and human health–a review. Environmental Health. 2006;5(1):1-23. DOI: 10.1186/1476-069X-5-7

[66] Larouche J, Sheoran S, Maruyama K, Martino MM. Immune regulation of skin wound healing: mechanisms and novel therapeutic targets. Advances in Wound Care. 2018;7(7):209-231. DOI: 10.1089/wound.2017.0761

[67] Hayashi O, Katoh T, Okuwaki Y. Enhancement of antibody production in mice by dietary *Spirulina platensis*. Journal of Nutritional Science and Vitaminology. 1994;40(5):431-441. DOI: 10.3177/jnsv.40.431

[68] Khan Z, Bhadouria P, Bisen PS. Nutritional and therapeutic potential of *Spirulina*. Current pharmaceutical biotechnology. 2005;6(5):373-379. DOI: 10.2174/138920105774370607

[69] Park HJ, Lee YJ, Ryu HK, Kim MH, Chung HW, Kim WY. A randomized double-blind, placebo-controlled study to establish the effects of *Spirulina* in elderly Koreans. Annals of Nutrition and Metabolism. 2008;52(4):322-328. DOI: 10.1159/000151486

[70] Punitha P, Bharathi V. In vitro anti-inflammatory activity in ethanolic extract of *Spirulina plantensis* and phytochemical analysis underneath GC-MS/MS. World Journal of

Pharmaceutical and Medical Research. 2017;3(11):217-225.

[71] Appel K, Munoz E, Navarrete C, Cruz-Teno C, Biller A, Thiemann E. Immunomodulatory and inhibitory effect of Immulina®, and Immunloges® in the Ig-E mediated activation of RBL-2H3 cells. A new role in allergic inflammatory responses. Plants. 2018;7(1):13. DOI: 10.3390/plants7010013

[72] Singh S, Kate BN, Banerjee UC. Bioactive compounds from cyanobacteria and microalgae: an overview. Critical reviews in biotechnology. 2005;25(3):73-95. DOI: 10.1080/07388550500248498

[73] Liao HF, Wu TJ, Tai JL, Chi MC, Lin LL. Immunomodulatory potential of the polysaccharide-rich extract from edible cyanobacterium Nostoc commune. Medical Sciences. 2015;3(4):112-123. Doi: 10.3390/medsci3040112

[74] Effmert U, Mundt S, Teuscher E. Investigations of the immunomodulatory effect of cyanobacterial extracts. Allergie und Immunologie. 1991;37(2):97-102.

[75] Shen PP, Zhao SW, Zheng WJ, Hua ZC, Shi Q, Liu ZT. Effects of cyanobacteria bloom extract on some parameters of immune function in mice. Toxicology letters. 2003;143(1):27-36. DOI: https://doi.org/10.1016/S0378-4274(03)00110-3

[76] Moosova Z, Pekarova M, Sindlerova LS, Vasicek O, Kubala L, Blaha L, Adamovsky O. Immunomodulatory effects of cyanobacterial toxin cylindro spermopsin on innate immune cells. Chemosphere. 2019;226:439-446. DOI: 10.1016/j.chemosphere.2019.03.143

[77] Silambarasan G, Ramanathan T, Nabeel MA, Kalaichelvan VK, Kathiresan K, Balasubramanian T. Anti-inflammatory activity of the marine cyanobacterium *Trichodesmium erythraeum* against carrageenan-induced paw oedema in wistar albino rats. European Journal of Inflammation. 2011;9(1):53-56. DOI: https://doi.org/10.1177/1721727X1100900108

[78] Krishnaveni J, Alarmal MS, Jency M. In vitro antiinflammatory activity of blue-green algae-*Geitlerinema splendidum* collected from Western Ghats, South India Asian Journal of Pharmaceutical and Clinical Research. 2018;11(5),407-409. DOI: https://doi.org/10.22159/ajpcr.2018.v11i5.25301

[79] Eming SA, Koch M, Krieger A, Brachvogel B, Kreft S, Bruckner-Tuderman L, Krieg T, Shannon JD, Fox JW. Differential proteomic analysis distinguishes tissue repair biomarker signatures in wound exudates obtained from normal healing and chronic wounds. Journal of proteome research. 2010;9(9):4758-4766. DOI: https://doi.org/10.1021/pr100456d

[80] Kapuścik A, Hrouzek P, Kuzma M, Bártová S, Novák P, Jokela J, Pflüger M, Eger A, Hundsberger H, Kopecký J. Novel Aeruginosin-865 from *Nostoc* sp. as a Potent Anti-inflammatory Agent. ChemBioChem. 2013;14(17):2329-2337. DOI: 10.1002/cbic.201300246

[81] Torres A, Enk CD, Hochberg M, Srebnik M. Porphyra-334, a potential natural source for UVA protective sunscreens. Photochemical & Photobiological Sciences. 2006;5(4):432-435. DOI: https://doi.org/10.1039/B517330M

[82] Singh SP, Klisch M, Sinha RP, Häder DP. Effects of abiotic stressors on synthesis of the mycosporine-like amino acid shinorine in the cyanobacterium *Anabaena variabilis* PCC 7937. Photochemistry and Photobiology.

2008;84(6):1500-1505. DOI: https://doi.org/10.1111/j.1751-1097.2008.00376.x

[83] Engene N, Choi H, Esquenazi E, Byrum T, Villa FA, Cao Z, Murray TF, Dorrestein PC, Gerwick L, Gerwick WH. Phylogeny-guided isolation of ethyl tumonoate A from the marine cyanobacterium cf. *Oscillatoria margaritifera*. Journal of natural products. 2011;74(8):1737-1743. DOI: https://doi.org/10.1021/np200236c

[84] Faltermann S, Hutter S, Christen V, Hettich T, Fent K. Anti-inflammatory activity of cyanobacterial serine protease inhibitors aeruginosin 828A and cyanopeptolin 1020 in human hepatoma cell line Huh7 and effects in Zebrafish (*Danio rerio*). Toxins. 2016;8(219):1-14.DOI: https://doi.org/10.3390/toxins8070219

[85] Motoyama K, Tanida Y, Hata K, Hayashi T, Hashim II, Higashi T, Ishitsuka Y, Kondo Y, Irie T, Kaneko S, Arima H. Anti-inflammatory effects of novel polysaccharide sacran extracted from cyanobacterium *Aphanothece sacrum* in various inflammatory animal models. Biological and Pharmaceutical Bulletin. 2016;39(7):1172-1178. DOI: 10.1248/bpb.b16-00208

[86] Bruno A, Rossi C, Marcolongo G, Di Lena A, Venzo A, Berrie CP, Corda D. Selective *in vivo* anti-inflammatory action of the galactolipid monogalactosyl diacylglycerol. European Journal of Pharmacology. 2005;524(1-3):159-168. DOI: 10.1016/j.ejphar.2005.09.023

[87] Stevenson CS, Capper EA, Roshak AK, Marquez B, Grace K, Gerwick WH, Jacobs RS, Marshall LA. Scytonemin-a marine natural product inhibitor of kinases key in hyperproliferative inflammatory diseases. Inflammation Research. 2002;51(2):112-114. DOI: 10.1007/BF02684014

[88] Balunas MJ, Grosso MF, Villa FA, Engene N, McPhail KL, Tidgewell K, Pineda LM, Gerwick L, Spadafora C, Kyle DE, Gerwick WH. Coibacins A–D, antileishmanial marine cyanobacterial polyketides with intriguing biosynthetic origins. Organic letters. 2012;14(15): 3878-3881. DOI: https://doi.org/10.1021/ol301607q

[89] Choi H, Mascuch SJ, Villa FA, Byrum T, Teasdale ME, Smith JE, Preskitt LB, Rowley DC, Gerwick L, Gerwick WH. Honaucins A– C, potent inhibitors of inflammation and bacterial quorum sensing: synthetic derivatives and structure-activity relationships. Chemistry & Biology. 2012;19(5):589-598. Doi: 10.1016/j.chembiol.2012.03.014

[90] Morrison CA. The prehospital treatment of the bleeding patient—Dare to dream. Journal of Surgical Research. 2013;180(2):246-247. DOI: 10.1016/j.jss.2011.12.022

[91] Dai C, Liu C, Wei J, Hong H, Zhao Q. Molecular imprinted macroporous chitosan coated mesoporous silica xerogels for hemorrhage control. Biomaterials. 2010;31(30):7620-7630. DOI: 10.1016/j.biomaterials.2010.06.049

[92] Choi JH, Kim S, Kim SJ. Spirulan from blue-green algae inhibits fibrin and blood clots: Its potent antithrombotic effects. Journal of Biochemical and Molecular Toxicology. 2015;29(5):240-248. DOI: https://doi.org/10.1002/jbt.21690

[93] Anas AR, Nakajima A, Naruse C, Tone M, Asukabe H, Harada KI. Determination of FVIIa-sTF Inhibitors in Toxic Microcystis Cyanobacteria by LC-MS Technique. Marine drugs. 2016;14(1):7.DOI: https://doi.org/10.3390/md14010007

[94] Jensen GS, Drapeau C, Lenninger M, Benson KF. Clinical safety of a high dose of Phycocyanin-enriched

aqueous extract from *Arthrospira (Spirulina) platensis*: Results from a randomized, double-blind, placebo-controlled study with a focus on anticoagulant activity and platelet activation. Journal of Medicinal Food. 2016;19(7):645-653. DOI: 10.1089/jmf.2015.0143

[95] Kamble SP, Gaikar RB, Padalia RB, Shinde KD. Extraction and purification of C-phycocyanin from dry *Spirulina* powder and evaluating its antioxidant, anticoagulation and prevention of DNA damage activity. Journal of Applied Pharmaceutical Science. 2013;3(8):149. DOI: 10.7324/JAPS.2013.3826

[96] Kaji T, Okabe M, Shimada S, Yamamoto C, Fujiwara Y, Lee JB, Hayashi T. Sodium spirulan as a potent inhibitor of arterial smooth muscle cell proliferation *in vitro*. Life Sciences. 2004 ;74(19):2431-2439. DOI: 10.1016/j.lfs.2003.09.061

[97] Harborne JB. Phytochemical methods; A guide to modern Techniques of plant analysis. Chapman and Hall. New York. 1998:1-98.

[98] Dahms HU, Ying X, Pfeiffer C. Antifouling potential of cyanobacteria: a mini-review. Biofouling. 2006;22(5):317-327. DOI: https://doi.org/10.1080/08927010600967261

[99] Kim HM, Lee EH, Cho HH, Moon YH. Inhibitory effect of mastcell-mediated immediate-type allergic reactions in rats by *Spirulina*, Biochemical Pharmacology 1998;55:1071-1076 DOI: 10.1016/s0006-2952(97)00678-3

[100] Subhashini J, Mahipal SV, Reddy MC, Reddy MM, Rachamallu A, Reddanna P. Molecular mechanisms in C-Phycocyanin induced apoptosis in human chronic myeloid leukemia cell line-K562. Biochemical pharmacology. 2004;68(3):453-462. DOI: 10.1016/j.bcp.2004.02.025

[101] Gur CS, Erdogan DK, Onbasılar I, Atilla P, Cakar N, Gurhan ID. In vitro and in vivo investigations of the wound healing effect of crude *Spirulina* extract and C-phycocyanin. Journal of Medicinal Plants Research. 2013;7(8):425-433. DOI: https://doi.org/10.5897/JMPR12.705

[102] Syarina PN, Karthivashan G, Abas F, Arulselvan P, Fakurazi S. Wound healing potential of *Spirulina platensis* extracts on human dermal fibroblast cells. EXCLI journal. 2015;14:385-393. DOI: 10.17179/excli2014-697

[103] Han SG, Song JM, Lee J. Formulating a therapeutic cream for burn injury using cyanobacterial extracts and gold nanoparticles. International Journal of Medical Research and Pharmaceutical Sciences. 2017;4(3):45-53. DOI: 10.5281/zenodo.376416

[104] Morais MG, Vaz BD, Morais EG, Costa JA. Biological effects of *Spirulina (Arthrospira)* biopolymers and biomass in the development of nanostructured scaffolds. BioMed Research International. 2014;2014. DOI: https://doi.org/10.1155/2014/762705

[105] Steffens D, Leonardi D, da Luz Soster PR, Lersch M, Rosa A, Crestani T, Scher C, de Morais MG, Costa JA, Pranke P. Development of a new nanofiber scaffold for use with stem cells in a third degree burn animal model. Burns. 2014;40(8):1650-1660. DOI: 10.1016/j.burns.2014.03.008

[106] Jung SM, Min SK, Lee HC, Kwon YS, Jung MH, Shin HS. *Spirulina*-PCL nanofiber wound dressing to improve cutaneous wound healing by enhancing antioxidative mechanism. Journal of Nanomaterials. 2016;2016 DOI: https://doi.org/10.1155/2016/6135727

[107] Yin H, Chen CY, Liu YW, Tan YJ, Deng ZL, Yang F, Huang FY, Wen C,

Rao SS, Luo MJ, Hu XK. *Synechococcus elongatus* PCC7942 secretes extracellular vesicles to accelerate cutaneous wound healing by promoting angiogenesis. Theranostics. 2019;9(9):2678-2693 DOI: 10.7150/thno.31884

[108] Liu W, Xu H, Gao X. Mild Acid Hydrolysis-related Release of Water-soluble Sunscreen Pigments from the Exopolysaccharide Matrix of Edible Terrestrial Cyanobacteria. Journal of Applied Biology & Biotechnology. 2017;5(5):11-15. DOI: 10.7324/JABB.2017.50502

[109] Choi YH, Yang DJ, Kulkarni A, Moh SH, Kim KW. Mycosporine-like amino acids promote wound healing through focal adhesion kinase (FAK) and mitogen-activated protein kinases (MAP kinases) signaling pathway in keratinocytes. Marine drugs. 2015;13(12):7055-7066. DOI: 10.3390/md13127056

[110] Peppas NA, Bures P, Leobandung WS, Ichikawa H. Hydrogels in pharmaceutical formulations. European Journal of Pharmaceutics and Biopharmaceutics. 2000;50(1):27-46. DOI: 10.1016/s0939-6411(00)00090-4

[111] Agnihotri SA, Mallikarjuna NN, Aminabhavi TM. Recent advances on chitosan-based micro-and nanoparticles in drug delivery. Journal of Controlled Release. 2004;100(1):5-28. DOI: 10.1016/j.jconrel.2004.08.010

[112] Oh JK, Lee DI, Park JM. Biopolymer-based microgels/nanogels for drug delivery applications. Progress in polymer science. 2009;34(12):1261-1282. DOI: 10.1016/j.progpolymsci.2009.08.001

Chapter 4

Novel Hemoglobin from *Synechocystis* sp. PCC 6803: Shedding Light on the Structure-Function Relationship and Its Biotechnological Applications

Mohd. Asim Khan, Sheetal Uppal and Suman Kundu

Abstract

Cyanobacteria are oxygenic photosynthetic prokaryotes, practically present in every plausible environment on the earth. In 1996, the first cyanobacterial genome was sequenced from *Synechocystis* sp. PCC 6803 and the cyanobacterial genome database has been continuously growing with genomes from more than 300 cyanobacterial and other related species, so far. *Synechocystis* sp. PCC 6803 is one of the best-characterized cyanobacteria and has developed into a model cyanobacterium that scientists are using throughout the world. At the same time, the field of hemoglobin was undergoing a breakthrough with the identification of new globins in all three kingdoms of life including cyanobacteria. Since then, the newly identified globins in the cyanobacteria are raising intriguing questions about their structure and physiological functions, which are quite different from vertebrate's hemoglobin and myoglobin. These hemoglobins have displayed unprecedented stability, unique heme coordination, novel conformational changes, and other properties that are not often observed in the globin superfamily. This chapter provides an overview of the unique globin from *Synechocystis sp.* PCC 6803, its interacting protein partners, proposed functions, and its biotechnological implications including potential in the field of artificial oxygen carriers.

Keywords: Cyanobacteria, *Synechocystis* sp. PCC 6803 hemoglobin, Structural features, Heme stability, Physiological function, Biotechnological application

1. Introduction

The ancient cyanobacteria played a fundamental role in changing the composition of the early, oxygen-poor reducing atmosphere into an oxidizing atmosphere of the earth. These tiny oxygenic phototrophs inhabit varied ecosystems and habitats ranging from oceans to hot springs and deserts [1]. They can also be found in extreme environments, such as acidic bogs and volcanoes. The plethora of available information on the diversity and physiology of cyanobacteria provides an excellent base for exploring their application in biotechnology. Because of their ability to harvest solar energy and convert atmospheric CO_2 to useful products like biofuels and bioactive compounds, they serve as a promising organism which is used for medical treatments and various industrial applications [2].

Oxygen provides an enormous source of energy for biological functions; however, it can also be toxic to organisms. It is believed that cyanobacteria were among the earliest prokaryotic organisms responsible for the oxygen-rich environment on the young planet earth and currently, nearly 99% of the oxygen is contributed by the eukaryotic algae [3]. It has been revealed that all the eukaryotic phototrophic organisms derived the ability to produce oxygen during photosynthesis through endosymbiosis [4]. Later, it was discovered that the heme-containing protein that sequesters and protects the primitive cyanobacteria cells from toxic O_2 is almost identical to the energy-generating apparatus in the photosynthetic bacteria [5]. The basic chemical apparatus became increasingly complex through time and evolution, however, the interaction between the metal atom in the porphyrin ring and the oxygen remains unchanged [6]. The heme-containing proteins form a large class of macromolecules that have diverse and distinct biological functions. Comparative analysis of hemoproteins revealed that the changes in the amino acid sequence and their interaction with the porphyrin ring mainly involved in the multitude of different functions which include electron-transferring cytochromes, intracellular peroxidases, and lignin-degrading extracellular peroxidases. The well-studied heme-binding proteins are vertebrate hemoglobin (Hb) and myoglobin (Mb). The heterotetrameric hemoglobin is present at high concentration (15 g/100 ml) in normal human blood and involved in the oxygen transport in the circulatory system whereas myoglobin is a monomeric oxygen storage protein mainly located in the cardiac and striated muscles [7].

Recent breakthroughs in molecular biology tools and genome sequencing techniques led to the identification of hemoglobin genes in almost all kingdoms of life including plants, animals, bacteria, and fungi. Extensive bioinformatics surveys of available genomes identified putative globins with several characteristics that are distinct from the classical globins [8]. These distinct globins have been designated as "novel" or "new" globins to distinguish them from the traditional globins. These new Hbs display differences in the coordination chemistry of heme Fe atom in which all six coordination sites are occupied in the absence of exogenous ligand and referred to as "hexacoordinated hemoglobins (HxHbs)" compared to "pentacoordinated" heme Fe coordination chemistry of classical vertebrate Hb and Mb [9, 10]. Another set of novel hemoglobins have been discovered which are 20–40 amino acid residues shorter than the mammalian hemoglobins, resulted in modification of the canonical "3-on-3" globin fold and provided them with a shortened "2-on-2" globin fold [11–13]. These classes of hemoglobins are called "truncated hemoglobins (TrHbs)" and constitute a major class of the globin family [14].

In 1992, the identification of truncated hemoglobin in the nitrogen (N_2)-fixing cyanobacterium *Nostoc commune (N. commune)* strain UTEX 584 opened the avenues for exploring and identification of hemoglobin gene in cyanobacteria and green algae [15]. Several years later, genome sequence analysis of the first photosynthetic non-N_2 fixing *Synechocystis* sp. PCC 6803 (S6803) revealed a single copy of the globin gene [16]. Among all cyanobacterial species, S6803 is one of the most widely studied cyanobacteria. In this article, we will provide an overview of the history of *Synechocystis* hemoglobin (*Syn*Hb), its proposed physiological functions and interacting partners as well as the biotechnological applications of the cyanobacterial hemoglobin in the designing of artificial oxygen carriers.

2. *Synechocystis* sp. PCC 6803 hemoglobin: an unusual hemoglobin from a cyanobacterium

S6803 was the third unicellular prokaryotic and first non-diazotrophic photosynthetic organism whose genome is completely sequenced. The genome sequence analysis showed the presence of a single hemoglobin gene (coded by slr2097 gene; named as glbN), encoding 123- amino acid polypeptide chain sharing 55% sequence identity with the cyanoglobin from *N. commune* [17]. Compared to *N. commune*, the location of the slr2097 gene in the genome does not provide any indication of a functional role for the protein. So, to uncover the physiological role, this cyanobacterial globin is being investigated worldwide and a series of research efforts by several pioneers provided the structural and functional understanding as outlined in **Figure 1**.

2.1 Biochemical and structural features of *Synechocystis* Hemoglobin

In the year 2000, two different groups reported the preliminary biochemical characterization by cloning and over-expressing the slr2097 globin gene in *E.coli* [17, 18, 32]. The molecular weight of *Syn*Hb protein is around ~13 kDa and was found to be a stable α-helical monomeric protein [33]. This cyanobacterial globin is "hexacoordinated" in which His46 (distal) and His70 (proximal) function as internal heme iron axial ligands [19]. Comparative sequence analysis of *Syn*Hb with pentacoordinated sperm whale Mb and other truncated hemoglobins (trHbs) showed that *Syn*Hb is a member of the truncated hemoglobin superfamily with "2-on-2" globin fold (**Figure 2**). The overall crystal structure of *Syn*Hb is almost similar to other trHbs and the NMR structure, however, some conformational changes were found, which was partly attributed to the presence of a unique third covalent linkage between the heme-2-vinyl and the Nε2 atom of His117 residue – a feature observed in crystal structure but not in the NMR structure. This unique post-translational modification with heme moiety is not detected in any other globins discovered so far. Interestingly, *Syn*Hb displays the characteristics features of both trHbs and HxHbs. It has been found that His46 residue covalently attached to the heme Fe atom on the distal side resulting in "hexacoordination". However, unlike in other HxHbs whose structures are solved, His46 occupies E10 position compared to common E7 position and not involved in the stabilization of the bound ligand (**Figure 3**). Interestingly, structural comparison of *Syn*Hb with other trHbs

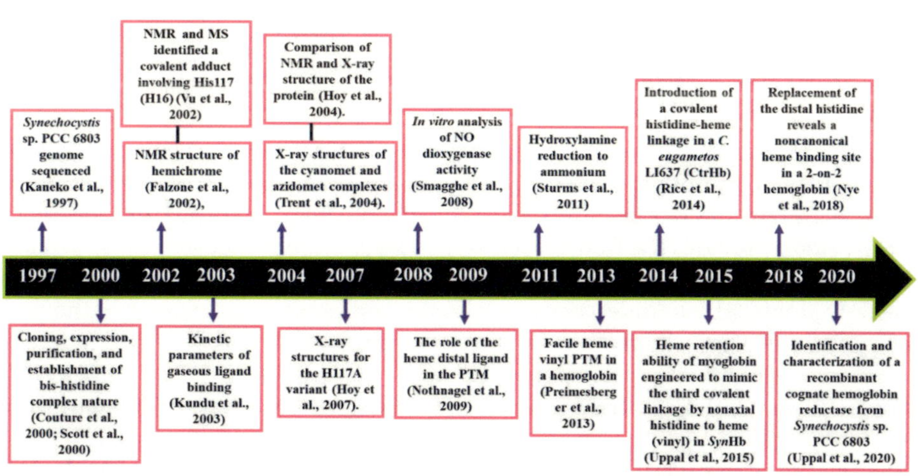

Figure 1.
Timeline of milestones in Synechocystis *hemoglobin research [16–31].*

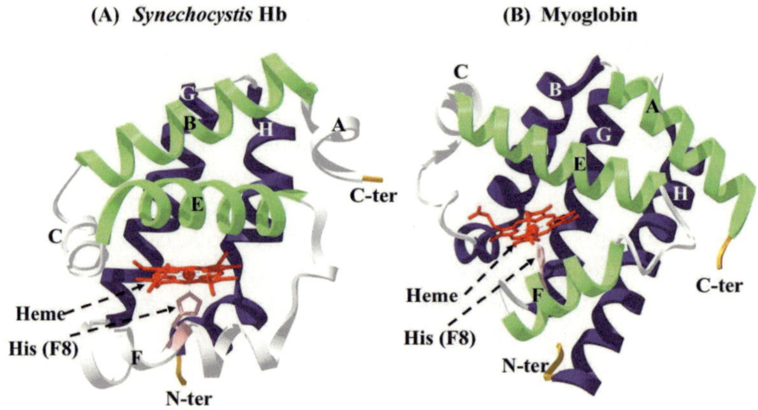

Figure 2.
Three-dimensional structure showing globin fold. A) a typical truncated Synechocystis *hemoglobin (PDB ID: 1RTX) showing "2-on-2" globin fold (BE (green)/GH (blue)). B) Three-dimensional structure of classical globin i.e. myoglobin (PDB ID: 5MBN) showing "3-on-3" globin fold (AEF (green)/BGH (blue)).*

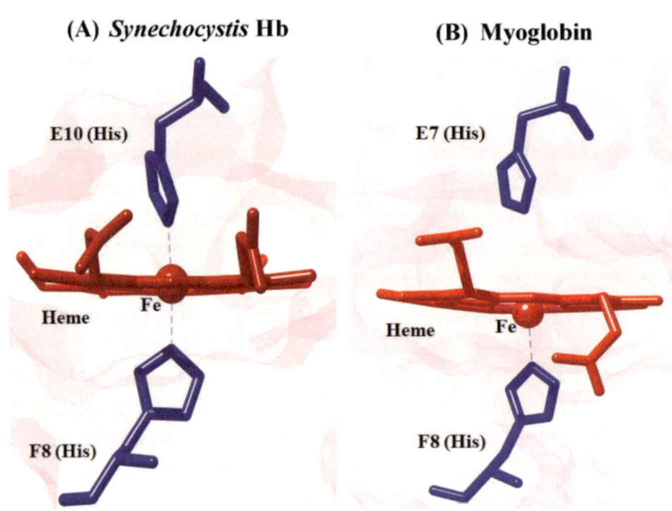

Figure 3.
A) Three dimensional structures of SynHb *(PDB-1RTX) shows hexacoordinate heme coordination chemistry with distal his (E10) and proximal his (F8) covalently linked to heme Fe atom. B) Three dimensional structures of Mb (PDB- 5MBN) shows pentacoordinate heme coordination chemistry with distal his (E7) not directly bound to heme iron atom whereas proximal his (F8) covalently linked to heme Fe atom.*

reveals the absence of a ligand tunnel, a characteristic feature of trHbs connecting the distal heme pocket to the solvent. The tunnel formation is observed post-ligand binding in *Syn*Hb and has been proposed to facilitate ligand escape from the heme pocket. *Syn*Hb provides the first crystal structure of a trHb in both unliganded and ligand-bound state [19, 34]. Among all the available hemoglobin structures, *Syn*Hb is the only globin which undergoes a significant conformational change in the tertiary structure upon ligand binding in hexacoordinated-*Syn*Hb [20].

2.2 Role of key residues in the stability and folding of *Synechocystis* Hemoglobin

Preliminary investigation of *Syn*Hb by Hoy et al. [35], and Nothnagel et al. [21], revealed that this cyanobacterial Hb is naturally a very stable Hb. Their work highlighted the role of thc unusual third His (His117) in imparting extra heme stability in *Syn*Hb. Though the globin has been reported to be stable, negligible reports exist as to its relative stability, extent of stability, the factors that contribute to it and their applications and thus *Synechocystis* hemoglobin is being investigated worldwide. In the past few years, Kundu's laboratory is intensively involved in studying this unique cyanobacterial globin to understand the biophysical traits that define this globin and structure–function relationship in comparison to classical Hbs. Extensive mutational studies and heme loss assay reiterated and validated some earlier propositions that the third covalent linkage between heme-2-vinyl group and His117 residue is the major holding force for heme in *Syn*Hb [22, 32]. Several key residues near the heme pocket have identified that influence the structural integrity of protein and thus play vital role in the cyanobacterial globin expression and synergy of amino acid side chains in the heme and polypeptide stability (**Figure 4**) [33]. Their studies have revealed several interesting findings which include: 1) His117 is indispensable for heme retention, while either of distal His46 or proximal His70 is required for heme uptake by apo-*Syn*Hb (globin protein without heme); 2) Acid-induced denaturation studies showed that *Syn*Hb did not release heme from its protein matrix and displayed features of a molten globule state at pH 2.0; 3) Acid- and chemical-induced denaturation studies revealed that none of the heme pocket residues affect the polypeptide stability except His117 residue; 4) This cyanobacterial Hb is extremely thermostable compared to classical pentacoordinated Hbs and thermal unfolding is affected both by distal His46 and His117 residues. Based on these findings, it has been proposed that *Synechocystis* Hb could be used as a model system for understanding the protein folding and stability of a new class of hemoglobins.

2.3 Ligand binding kinetics in *Synechocystis* Hb

The reaction of *Syn*Hb with O_2 and CO is quite unusual. The association rate constants for O_2 and CO is partly similar to the hemoprotein (glbN) from the *N. commune* cyanobacterium [36]. Compared to other hemoglobins (**Table 1**), such as Mb and human Hb, soybean leghemoglobin (Lba), and non-symbiotic hemoglobin (nsHb) from rice, these two cyanobacterial proteins have remarkably large values for most rate constants [38–42]. The two cyanobacterial Hbs are different in their O_2

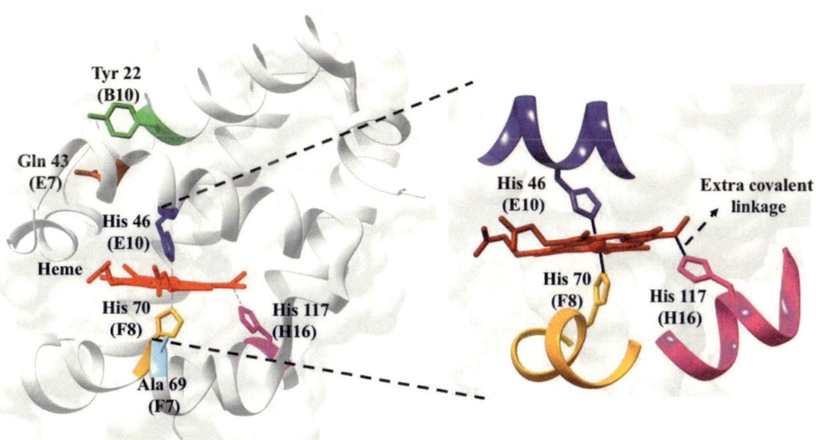

Figure 4.
Structural representation of Synechocystis *hemoglobin (PDB ID: 1RTX). The truncated globin fold (2-on-2) is represented in gray and the heme moiety in red. Several key residues in the heme pocket are displayed in color. The three histidines covalently associated with heme are shown in zoomed section. His46 and His70 directly coordinate to heme iron and constitute "hexacoordination". The third His (His117) is covalently associated to heme vinyl group.*

Hb	k'_{O2} ($\mu M^{-1} s^{-1}$)	k_{O2} (s^{-1})	k'_{CO2} ($\mu M^{-1} s^{-1}$)	$k_{entry, CO}$ ($\mu M^{-1} s^{-1}$)
Synechocystis hyobin	240	0.014	90	
Nostoc commune (glbN)	390	79	41	
M. tuberculosis Hb (HbN)	25	0.2	6.8	
Rice Hb (rHb1)	68	0.038	6	
Sperm whale Mb	17	15	0.5	17
Human Hb α- chain	23	11	2.9	11
Human Hb β- chain	79	28	7.1	11
Soybean LegHb (Lba)	130	5.6	17	320

k_{O2} = association rate constant (on-rate); k_{O2} = dissociation rate constant (off-rate); k'_{CO2} = association rate constant (on-rate); $k_{entry,CO}$ = rate constant for entry through the protein matrix.

Table 1.
Rate constants for ligand binding to several wild type hemoglobins [21, 37].

dissociation rate constants. The oxygen (O_2) dissociation rate constant of *Syn*Hb is very low compared to faster O_2 dissociation rate of glbN (**Table 1**). The low dissociation constant of *Syn*Hb is similar to other HxHbs like rice Hb1. All classical Hbs have a relatively higher off rate. *Syn*Hb exhibits large association rate constants for CO and O_2 which is because of unusual reactive heme iron, suggesting that the ligand is trapped in the heme pocket which increases the chance of bond formation rather than escape (**Table 1**). Thus, bimolecular rate constants cannot surpass the rate constant for ligand entry through the protein matrix. It was observed that k'_{CO} for *Syn*Hb is larger than $k_{entry,CO}$ for Mb and human Hb; thus indicating that the heme pocket of *Syn*Hb is highly solvent-exposed. The unusual ligand binding kinetics in *Syn*Hb may not support a role in O_2 transport and storage. In trHbs, the conventional "histidine-gate" path for ligand binding is blocked [43]. It has been postulated that tunnels found in trHbs serve as an alternative diffusion path for ligands. In case of *Syn*Hb, there is no tunnel observed and subsequently, three possibilities are proposed for ligand entry and exit which includes: 1) Ligands enter and exit directly from the heme pocket through the solvent face. 2) Ligands enter through the solvent face of the heme pocket, and then a tunnel formation in the ligand bound state facilitate the ligand exit. 3) Ligand entry and exit pathways are not clear in *Syn*Hb structures.

3. Proposed physiological function of *Synechocystis* hemoglobin

The new class of Hb differs from the classical Hbs in their cellular location, primary sequence, expression pattern, three-dimensional fold, heme pocket architecture, ligand binding characteristics, heme pocket electrostatics etc. and thus, forced the researchers worldwide to re-investigate their functions which might diverge widely during evolution of globins [11, 44]. Based on several reports, the oxygen transport and storage function which are usually associated with hemoglobins have been ruled out and various other functions including detection, scavenging, and detoxification of O_2 and O_2- derived species (e.g. NO and CO) have been proposed.

Despite numerous efforts, physiological function for *Syn*Hb is still not clear. Several functions have been proposed that vary from reversible binding of diatomic ligands to redox reactions, to peroxidase activity, and nitrite and hydroxylamine chemistry [23, 24, 45–48]. The possibility for *Syn*Hb to function as an oxygen transporter protein is excluded as the oxygen dissociation rate constant of *Syn*Hb is too low for its involvement in the facilitated diffusion of oxygen, thus suggesting that it may be involved in

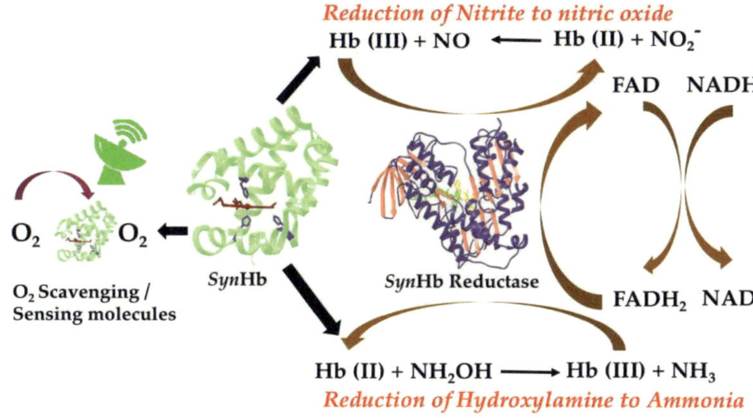

Figure 5.
Proposed working model of Synechocystis hemoglobin-reductase complex.

O_2 scavenging or sensing mechanism [25, 37]. In cyanobacteria, the metabolism of N_2 is different and assimilates nitrogen through the reduction of nitrate under conditions ranging from normoxic to hypoxic [24]. It has been shown that at low oxygen concentration, cyanobacteria can accumulate nitrate and nitrite to a very high concentration in the range of millimolar [49]. Sturms et al., demonstrated that the ferrous *Syn*Hb showed nitrite reduction rate 10 times faster than the animal hemoglobins [24], which indicates that cyanobacterial Hb could serve as an anaerobic nitrite reductase *in vivo*. Similarly, the same group has shown that the hydroxylamine reduction rate to ammonia is 100–2500 times faster than animal hemoglobins *in vitro* [23], which supports the hypothesis that it contribute to anaerobic nitrogen metabolism in support of anaerobic respiration and survival during hypoxia. Another existing study suggested that *Syn*Hb can serve as nitric oxide dioxygenase (NOD) enzyme by substituting for flavohemoglobins (FHbs) in *E. coli* during NO challenge [26]. In an artificial reduction system, oxyferrous-*Syn*Hb can react rapidly with NO and subsequently scavenge NO with a controlled reduction rate *in vitro*. Interestingly, it has been in *Synechococcous* sp. PCC 7002, the close relative of *Synechocystis* sp. PCC 6803, that globin null mutant strain experienced more stress under NO exposure compared to wild type strain [50]. These NO protection/scavenging functions would thus need the requirement of a suitable reduction mechanism that converts inactive ferric state of *Syn*Hb back to active ferrous state of *Syn*Hb to accomplish these functions *in vivo*. These studies prompted other researchers to search for related reductase protein for *Syn*Hb which can perform the proposed nitrite reductase function. Recently, Uppal et al., reported a protein in *Synechocystis* sp. PCC 6803 which was annotated as dihydrolipoamide dehydrogenase (DLDH) in the database. Their studies clearly showed that the putative cognate reductase "named as *Synechocystis* Hb reductase (*Syn*HbRed)" interact and reduce the Fe^{3+}-*Syn*Hb back to active Fe^{2+}-*Syn*Hb *in vitro* [51]. Based on these results, we proposed a hypothetical *Synechocystis* hemoglobin-reductase reduction system as shown in **Figure 5**.

4. Biotechnological applications of *Synechocystis* hemoglobin

4.1 In designing a stable hemoglobin-based blood substitute using protein engineering approach

In the last few decades, there is a significant progress in the development of oxygen-carrying blood substitutes. Since the 1980s, human blood substitutes have been in the pipeline in the medical and life science research fields [52]. Currently, there are none in the market because of scientific and political reasons. There are a few blood substitutes still progressing through clinical trials, and the academic community is still actively improving the products, also known as oxygen therapeutics and hemoglobin-based oxygen carriers [53, 54]. Over the last few years, studies have focused on developing "recombinant hemoglobin-based oxygen carriers" (rHBOCs) which can be used as an alternative to blood during transfusion therapy. Recombinant human hemoglobin is produced in heterologous expression systems like *E. coli* to fulfill the need for artificial blood substitutes [55–59]. However, such hemoglobins suffer quite a few disadvantages like dissociation into dimers, poor stability, easy clearance from circulation, high blood pressure, poor expression yields, improper ligand affinities and fast heme dissociation [60]. Recently, protein engineering approaches have been employed for designing more stable Hb-based blood substitutes, with several properties improved. Several laboratories are intensively involved to tackle the major remaining problems associated with artificial blood substitutes like stability and rapid heme dissociation [61]. Mb has been invariably used as a model protein for the commercial development of blood substitutes [62]. This small, stable, and well-studied Hb thus became a gold standard for protein engineering approaches.

The newly discovered truncated and hexacoordinate globins exhibit unique features that allow the exploration of a whole range of proteins, some of which might be more stable than Mb, thus allowing newer ways for comparative mutagenesis strategies to improve stability. *Synechocystis* hemoglobin with its unique His-heme linkage and unparalleled stability serves as an excellent reference system. Thus, the major application of *Synechocystis* hemoglobin is in the designing of stable hemoglobin-based blood substitutes, an area of translational science in hemoglobin biotechnology. It has been assumed that introduction of extra covalent linkage via histidine to heme in other globin might allow a new strategy to enhance heme stability. Lecomte's group was successful in engineering covalent linkage itself by substituting Leu79 to His and demonstrated that single variant L79H/H117A bound the heme weakly but nonetheless formed a covalent linkage between His79Nε2-heme 2-vinyl atom, analogous to His117-heme 2-vinyl linkage [27]. Another successful attempt to engineer the heme stability was done by Uppal et al., in myoglobin as a first step toward the production of a stable hemoglobin-based oxygen carriers [22]. Their work clearly demonstrated that the Mb mutant (I107H) with the engineered covalent linkage holds heme tightly, stable to denaturants and exhibited ligand binding kinetics similar to wild-type protein. The future perspective of this work is to engineer the extra covalent linkage in recombinant human Hb, necessitating a step toward the production of stable hemoglobin-based oxygen carriers.

4.2 *Synechocystis* hemoglobin as a fusion tag for enhancing the expression, solubility and purification of other proteins

Recent years have witnessed tremendous increase in the number of tags and the development of fusion strategies to facilitate the expression, purification, and solubilization of recombinant proteins which can be used as industrial enzymes, for drug discovery, and biotherapeutics [63]. There are now a wide variety of fusion tags available in the market which are well-characterized and used in the biotechnological industry to obtain highly purified biologically active recombinant proteins [64]. However, these available tags have a major limitation, i.e., the absence of any color to facilitate the visualization of target protein during the expression and purification process. Hemoglobins because of their distinctive bright red color, high solubility and stability offer a unique advantage of tracking of the target fused protein during expression and at different steps of purification and even in crystallization and thus minimizing the cost and time in fusion protein technology. Previous reports showed the use of visible tag systems such as flavoenzymes and hemeproteins that contain colored chromophores [65]. The *Vitreoscilla* hemoglobin (VHb) from the bacterium *Vitreoscilla* has been used successfully as a fusion expression vector for the production of many target proteins [66]. Based on recent biochemical studies of newly discovered hemoglobins, it is revealed that *Synechocystis* hemoglobin is the well-suited protein for high-throughput protein expression and purification [67]. Since it is small, very stable, highly soluble with a high expression yield, it can improve the expression yield and solubility of the desired protein.

5. Conclusion

In the last few decades, an intense research effort from several researchers worldwide has enabled us to uncover the unique properties of *Synechocystis* hemoglobin and shed the light on its physiological function. Despite extensive information available for *Syn*Hb, there are still many unanswered questions that need to be investigated. Moreover, the evolutionary significance of the third His-heme covalent linkage and its role in the physiological function of *Syn*Hb is still unclear. However, *Synechocystis* Hb presents an important model system to understand protein folding and stability in general. Furthermore, the knowledge gained from mutational and expression studies in *Synechocystis* Hb can be applied to other globins e.g. human hemoglobin for designing an efficient oxygen delivery vehicle.

Acknowledgements

Suman Kundu acknowledge the financial assistance from Life Science Research Board, Defense Research and Development Organization (LSRB-DRDO; Grant ID O/o DG (TM)/81/48222/LSRB317/SH&DD/2017) New Delhi, India, University of Delhi (R&D grant), Department of Science and Technology (PURSE), Government of India and University Grants Commission (SAP grant), and Department of Biotechnology, Government of India. Mohd. Asim Khan acknowledges LSRB-DRDO, for Senior Research Fellowship (SRF). Sheetal Uppal is thankful to National Eye Institute, NIH for providing financial assistance.

Conflict of interest

The authors declare no conflict of interest.

Abbreviations

GlbN	Cyanoglobin
HxHbs	Hexacoordinate hemoglobin
FHb	Flavohemoglobin
Hb	Hemoglobin
*Syn*Hb	*Synechocystis* hemoglobin
NMR	Nuclear magnetic resonance
NOD	Nitric oxide dioxygenase
nsHb	Non-symbiotic hemoglobin

PDB	Protein data bank
TrHb	Truncated hemoglobin
O_2	Oxygen
CO	Carbon monoxide
NO	Nitric oxide
HBOCs	Hemoglobin based oxygen carriers

Author details

Mohd. Asim Khan[1], Sheetal Uppal[2] and Suman Kundu[1*]

1 Department of Biochemistry, University of Delhi South Campus, New Delhi, India

2 National Eye Institute, National Institutes of Health, Bethesda, MD, USA

*Address all correspondence to: suman.kundu@south.du.ac.in

References

[1] Pisciotta JM, Zou Y, Baskakov IV. Light-dependent electrogenic activity of cyanobacteria. PloS one. 2010;5(5): e10821.

[2] Singh S, Kate BN, Banerjee UC. Bioactive compounds from cyanobacteria and microalgae: an overview. Critical reviews in biotechnology. 2005;25(3):73-95.

[3] Kasting JF, Siefert JL. Life and the evolution of Earth's atmosphere. Science. 2002;296(5570):1066-1068.

[4] Margulis L. Symbiosis in cell evolution : microbial communities in the Archean and Proterozoic eons. San Francisco, 2 ed: New York (N.Y.) : Freeman; 1993.

[5] Goldhor S. The History of Cell Respiration and Cytochrome. The Yale Journal of Biology and Medicine. 1966;39(3):207.

[6] Hardison R. The evolution of hemoglobin: studies of a very ancient protein suggest that changes in gene regulation are an important part of the evolutionary story. American Scientist. 1999;87(2):126-137.

[7] Hardison R. Hemoglobins from bacteria to man: evolution of different patterns of gene expression. Journal of Experimental Biology. 1998;201(8):1099-1117.

[8] Vinogradov SN, Tinajero-Trejo M, Poole RK, Hoogewijs D. Bacterial and archaeal globins - a revised perspective. Biochimica et Biophysica Acta (BBA)-Proteins and Proteomics. 2013;1834(9):1789-800.

[9] Burmester T, Ebner B, Weich B, Hankeln T. Cytoglobin: a novel globin type ubiquitously expressed invertebrate tissues. Molecular biology and evolution. 2002;19(4):416-421.

[10] Trent III JT, Watts RA, Hargrove MS. Human neuroglobin, a hexacoordinate hemoglobin that reversibly binds oxygen. Journal of Biological Chemistry. 2001; 276(32):30106-30110.

[11] Milani M, Pesce A, Ouellet Y, Ascenzi P, Guertin M, Bolognesi M. Mycobacterium tuberculosis hemoglobin N displays a protein tunnel suited for O2 diffusion to the heme. The EMBO journal. 2001;20(15):3902-3909.

[12] Moens L, Vanfleteren J, Van de Peer Y, Peeters K, Kapp O, Czeluzniak J, et al. Globins in nonvertebrate species: dispersal by horizontal gene transfer and evolution of the structure-function relationships. Molecular biology and evolution. 1996;13(2):324-333.

[13] Pesce A, Couture M, Dewilde S, Guertin M, Yamauchi K, Ascenzi P, et al. A novel two-over-two α-helical sandwich fold is characteristic of the truncated hemoglobin family. The EMBO journal. 2000;19(11):2424-2434.

[14] Kumar A, Nag M, Basak S. Truncated or 2/2 hemoglobins: a new class of globins with novel structure and function. Journal of proteins and proteomics. 2013;4(1):45-64.

[15] Potts M, Angeloni SV, Ebel RE, Bassam D. Myoglobin in a cyanobacterium. Science. 1992;256(5064):1690-1692.

[16] Kaneko T, Tabata S. Complete genome structure of the unicellular cyanobacterium Synechocystis sp. PCC6803. Plant and Cell Physiology. 1997;38(11):1171-1176.

[17] Scott NL, Lecomte JT. Cloning, expression, purification, and preliminary characterization of a putative hemoglobin from the cyanobacterium Synechocystis sp. PCC

6803. Protein Science. 2000;9(3):587-597.

[18] Couture M, Das TK, Savard PY, Ouellet Y, Wittenberg JB, Wittenberg BA, et al. Structural investigations of the hemoglobin of the cyanobacterium Synechocystis PCC6803 reveal a unique distal heme pocket. European journal of biochemistry. 2000;267(15):4770-4780.

[19] Hoy JA, Kundu S, Trent JT, Ramaswamy S, Hargrove MS. The crystal structure of Synechocystis hemoglobin with a covalent heme linkage. Journal of Biological Chemistry. 2004;279(16):16535-16542.

[20] Trent III JT, Kundu S, Hoy JA, Hargrove MS. Crystallographic analysis of Synechocystis cyanoglobin reveals the structural changes accompanying ligand binding in a hexacoordinate hemoglobin. Journal of molecular biology. 2004;341(4):1097-1108.

[21] Nothnagel HJ, Love N, Lecomte JTJ. The role of the heme distal ligand in the post-translational modification of Synechocystis hemoglobin. Journal of inorganic biochemistry. 2009;103(1):107-116.

[22] Uppal S, Salhotra S, Mukhi N, Zaidi FK, Seal M, Dey SG, et al. Significantly enhanced heme retention ability of myoglobin engineered to mimic the third covalent linkage by nonaxial histidine to heme (vinyl) in synechocystis hemoglobin. Journal of Biological Chemistry. 2015;290(4):1979-1993.

[23] Sturms R, DiSpirito AA, Fulton DB, Hargrove MS. Hydroxylamine reduction to ammonium by plant and cyanobacterial hemoglobins. Biochemistry. 2011;50(50):10829-10835.

[24] Sturms R, DiSpirito AA, Hargrove MS. Plant and cyanobacterial hemoglobins reduce nitrite to nitric oxide under anoxic conditions. Biochemistry. 2011;50(19):3873-3878.

[25] Kundu S, Premer SA, Hoy JA, Trent III JT, Hargrove MS. Direct measurement of equilibrium constants for high-affinity hemoglobins. Biophysical Journal. 2003;84(6): 3931-3940.

[26] Smagghe BJ, Trent JT, 3rd, Hargrove MS. NO dioxygenase activity in hemoglobins is ubiquitous in vitro, but limited by reduction in vivo. PLoS One. 2008;3(4):e2039.

[27] Preimesberger MR, Wenke BB, Gilevicius L, Pond MP, Lecomte JTJ. Facile heme vinyl posttranslational modification in a hemoglobin. Biochemistry. 2013;52(20):3478-3488.

[28] Falzone CJ, Christie Vu B, Scott NL, Lecomte JT. The solution structure of the recombinant hemoglobin from the cyanobacterium Synechocystis sp. PCC 6803 in its hemichrome state. J Mol Biol. 2002;324(5):1015-1029.

[29] Vu BC, Jones AD, Lecomte JT. Novel histidine-heme covalent linkage in a hemoglobin. J Am Chem Soc. 2002;124(29):8544-8545.

[30] Hoy JA, Smagghe BJ, Halder P, Hargrove MS. Covalent heme attachment in Synechocystis hemoglobin is required to prevent ferrous heme dissociation. Protein Science. 2007;16(2):250-260.

[31] Nye DB, Lecomte JTJ. Replacement of the Distal Histidine Reveals a Noncanonical Heme Binding Site in a 2-on-2 Hemoglobin. Biochemistry. 2018;57(40):5785-5796.

[32] Lecomte JTJ, Scott NL, Vu BC, Falzone CJ. Binding of ferric heme by the recombinant globin from the cyanobacterium Synechocystis sp. PCC 6803. Biochemistry. 2001;40(21): 6541-6552.

[33] Uppal S, Khan MA, Kundu S. Stability and Folding of the Unusually Stable Hemoglobin from Synechocystis is Subtly Optimized and Dependent on the Key Heme Pocket Residues. Protein and Peptide Letters. 2020.

[34] Samuni U, Dantsker D, Ray A, Wittenberg JB, Wittenberg BA, Dewilde S, et al. Kinetic Modulation in Carbonmonoxy Derivatives of Truncated Hemoglobins THE ROLE OF DISTAL HEME POCKET RESIDUES AND EXTENDED APOLAR TUNNEL. Journal of Biological Chemistry. 2003;278(29):27241-27250.

[35] Hoy JA, Smagghe BJ, Halder P, Hargrove MS. Covalent heme attachment in Synechocystis hemoglobin is required to prevent ferrous heme dissociation. Protein Sci. 2007;16(2):250-260.

[36] Thorsteinsson MV, Bevan DR, Potts M, Dou Y, Eich RF, Hargrove MS, et al. A cyanobacterial hemoglobin with unusual ligand binding kinetics and stability properties. Biochemistry. 1999;38(7):2117-2126.

[37] Hvitved AN, Trent III JT, Premer SA, Hargrove MS. Ligand binding and hexacoordination in synechocystishemoglobin. Journal of Biological Chemistry. 2001;276(37): 34714-34721.

[38] Goodman MD, Hargrove MS. Quaternary structure of rice nonsymbiotic hemoglobin. Journal of Biological Chemistry. 2001;276(9):6834-6839.

[39] Hargrove MS. A flash photolysis method to characterize hexacoordinate hemoglobin kinetics. Biophysical Journal. 2000;79(5):2733-2738.

[40] Hargrove MS, Barry JK, Brucker EA, Berry MB, Phillips Jr GN, Olson JS, et al. Characterization of recombinant soybean leghemoglobin a and apolar distal histidine mutants. Journal of molecular biology. 1997;266(5): 1032-1042.

[41] Olson JS, Mathews AJ, Rohlfs RJ, Springer BA, Egeberg KD, Sligar SG, et al. The role of the distal histidine in myoglobin and haemoglobin. Nature. 1988;336(6196):265-266.

[42] Rohlfs RJ, Olson JS, Gibson QH. A comparison of the geminate recombination kinetics of several monomeric heme proteins. Journal of Biological Chemistry. 1988;263(4):1803-1813.

[43] Halder P, Trent Iii JT, Hargrove MS. Influence of the protein matrix on intramolecular histidine ligation in ferric and ferrous hexacoordinate hemoglobins. PROTEINS: Structure, Function, and Bioinformatics. 2007;66(1):172-182.

[44] Vinogradov SN, Moens L. Diversity of globin function: enzymatic, transport, storage, and sensing. Journal of Biological Chemistry. 2008;283(14):8773-8777.

[45] Kundu S, Trent Iii JT, Hargrove MS. Plants, humans and hemoglobins. Trends in plant science. 2003;8(8):387-393.

[46] Olson JS, Phillips Jr GN. Myoglobin discriminates between O 2, NO, and CO by electrostatic interactions with the bound ligand. JBIC Journal of Biological Inorganic Chemistry. 1997;2(4):544-552.

[47] Kawada N, Kristensen DB, Asahina K, Nakatani K, Minamiyama Y, Seki S, et al. Characterization of a stellate cell activation-associated protein (STAP) with peroxidase activity found in rat hepatic stellate cells. Journal of Biological Chemistry. 2001;276(27):25318-25323.

[48] Sakamoto A, Sakurao S-h, Fukunaga K, Matsubara T,

Ueda-Hashimoto M, Tsukamoto S, et al. Three distinct Arabidopsis hemoglobins exhibit peroxidase-like activity and differentially mediate nitrite-dependent protein nitration. FEBS letters. 2004;572(1-3):27-32.

[49] Gupta KJ, Fernie AR, Kaiser WM, van Dongen JT. On the origins of nitric oxide. Trends in plant science. 2011;16(3):160-168.

[50] Scott NL, Xu Y, Shen G, Vuletich DA, Falzone CJ, Li Z, et al. Functional and structural characterization of the 2/2 hemoglobin from Synechococcus sp. PCC 7002. Biochemistry. 2010;49(33):7000-7011.

[51] Uppal S, Khan MA, Kundu S. Identification and characterization of a recombinant cognate hemoglobin reductase from Synechocystis sp. PCC 6803. International Journal of Biological Macromolecules. 2020;162:1054-1063.

[52] Winslow RM. Blood substitutes--a moving target. Nature medicine. 1995;1(11):1212-1215.

[53] Winslow RM. Current status of oxygen carriers ('blood substitutes'): 2006. Vox sanguinis. 2006;91(2):102-110.

[54] Alayash AI. Hemoglobin-based blood substitutes and the treatment of sickle cell disease: more harm than help? Biomolecules. 2017;7(1):2.

[55] Varadarajan R, Szabo A, Boxer SG. Cloning, expression in Escherichia coli, and reconstitution of human myoglobin. Proceedings of the National Academy of Sciences. 1985;82(17): 5681-5684.

[56] Springer BA, Sligar SG. High-level expression of sperm whale myoglobin in Escherichia coli. Proceedings of the National Academy of Sciences. 1987;84(24):8961-8965.

[57] Hoffman SJ, Looker DL, Roehrich JM, Cozart PE, Durfee SL, Tedesco JL, et al. Expression of fully functional tetrameric human hemoglobin in Escherichia coli. Proceedings of the National Academy of Sciences. 1990;87(21): 8521-8525.

[58] Looker D, Abbott-Brown D, Cozart P, Durfee S, Hoffman S, Mathews AJ, et al. A human recombinant haemoglobin designed for use as a blood substitute. Nature. 1992;356(6366):258-260.

[59] Looker D, Mathews AJ, Neway JO, Stetler GL. Expression of recombinant human hemoglobin in *Escherichia coli*. Methods in enzymology. 231: Elsevier; 1994. p. 364-74.

[60] Varnado CL, Mollan TL, Birukou I, Smith BJZ, Henderson DP, Olson JS. Development of recombinant hemoglobin-based oxygen carriers. Antioxidants & redox signaling. 2013;18(17):2314-2328.

[61] Olson JS, Eich RF, Smith LP, Warren JJ, Knowles BC. Protein engineering strategies for designing more stable hemoglobin-based blood substitutes. Artificial Cells, Blood Substitutes, and Biotechnology. 1997;25(1-2):227-241.

[62] Dou Y, Maillett DH, Eich RF, Olson JS. Myoglobin as a model system for designing heme protein based blood substitutes. Biophysical chemistry. 2002;98(1-2):127-148.

[63] Ki MR, Pack SP. Fusion tags to enhance heterologous protein expression. Appl Microbiol Biotechnol. 2020;104(6):2411-2425.

[64] Kosobokova EN, Skrypnik KA, Kosorukov VS. Overview of Fusion Tags for Recombinant Proteins. Biochemistry (Mosc). 2016;81(3):187-200.

[65] Robert D. Finn, Iouri Kapelioukh, Paine MJI. Rainbow tags: a visual tag system for recombinant protein expression and purification. Biotechniques. 2005;38(3):387-392.

[66] Kwon SY, Choi YJ, Kang TH, Lee KH, Cha SS, Kim GH, et al. Highly efficient protein expression and purification using bacterial hemoglobin fusion vector. Plasmid. 2005;53(3):274-282.

[67] Kumar A, Uppal S, Kundu S. The red goldmine: promises of biotechnological riches. 2009. In: Biotechnology Applications [Internet]. I. K. Publishing House, New Delhi, India [230-62].

Chapter 5

Brazilian Coast: A Significant Gap in the Knowledge of Cyanobacteria and Their Applications

Taiara A. Caires and Helen Michelle de J. Affe

Abstract

Brazil has 10.959 km of coastline which includes three ecoregions based on the biogeographic system, exhibiting a wide range of environments that favor the occurrence of numerous cyanobacterial morpho- and ecotypes. These organisms have a great adaptive capacity, which explains their occupancy in numerous environments and the high diversification of the group. Historically, the cyanobacteria have been classified only based on morphology, which makes their taxonomy quite challenging. There is usually little morphological variation between taxa, which makes it difficult to identify diacritical characteristics between some genera and species, making intergeneric and intraspecific delimitation tough. Thereby, the polyphasic approach based on different tools allows the identification of new taxa and the reassessment of those already established with more reliability, contributing to a better systematic resolution of the world 'cyanoflora', a term that we propose herein to describe the diversity of Cyanobacteria into Phycoflora area. However, the use of these tools is still not widely applied to most genera and species, especially those from tropical and subtropical environments, which has limited the real recognition of their biodiversity, as well as the knowledge about the cyanobacteria's evolutionary history and biogeography. In Brazil, even with the great development of phycological studies, the knowledge about Cyanobacteria from marine benthic environments has not evolved to the same degree. This phylum has been neglected in floristic surveys, presenting only 46 benthic species reported to the long Brazilian coastline, evidencing the still incipient knowledge about the diversity and distribution of this microorganism's group. Furthermore, biotechnological properties of Brazilian marine cyanobacteria are still almost completely unknown, with only three studies carried out to date, underestimating one of the most diverse groups and with promising potential for the possibility of isolating new biochemically active compounds. The ten new taxa related to the Brazilian coast in the last decade emphasizes the challenge of conducting further floristic surveys in the underexplored marine environments in order to fill an important lacune in the cyanoflora knowledge, as well as their biogeographic distribution and biotechnological potential. Besides, the recognition of the Brazilian cyanoflora makes an important contribution to the understanding of the functioning and monitoring of marine ecosystems and provide data for the construction of future public policies, which is a goal of the United Nations Decade for Ocean Science for Sustainable Development.

Keywords: biodiversity, biotechnological prospection, Brazilian coast, cyanoflora, cyanobacteria, polyphasic approach, taxonomy

1. Introduction

The phylum Cyanobacteria is a well-defined group of gram-negative oxygenic photosynthetic bacteria, responsible for the initial change of the Earth's atmosphere from the reducing to the oxidizing condition [1–4]. The origin of this group is estimated between 2.7 to 3.5 billion years, based on fossil records and organic biomarkers. Cyanobacteria are highly adaptable and have colonized the most varied biotopes as terrestrial, freshwater, brackish, and marine environments, including those considered extreme, such as the poles, hot springs, and deserts, being considered excellent environmental colonizers [5–7].

The adaptive capacity and diversification of cyanobacteria, as well as the occupation of the environments by these microorganisms can be explained by their great flexibility to acclimatize to a wide range of environmental conditions [1, 4, 8]. This extraordinary adaptability is favored by a wide variety of morphological and physiological characteristics exhibited by different eco- and morphotypes into Cyanobacteria [7]. Besides their significant ecological importance as primary producers, these organisms also are recognized as great fixers of atmospheric nitrogen, playing an important role in the bioavailability of the nitrogen compounds in the trophic chains of the most diverse environments, whether aquatic or terrestrial [9].

Cyanobacteria usually present phenotypic plasticity among the individuals of the same species. On the other hand, individuals from different species can represent cryptic taxa [10]. For marine environments, these morphological variations may be due to local environmental features, such as hydrodynamics, water temperature, shading, and substrate types [11]. According to Dvořák et al. [12], the difficulty of cyanobacteria identification stems from their cryptic diversity, as well as the poor knowledge about morphological variability of these organisms and the common convergent evolutionary events. Thus, the polymorphism presented by cyanobacteria makes it difficult to identify diacritical characteristics for species identification [11, 13].

Historically, the cyanobacteria have been classified only based on morphology, which makes their taxonomy quite challenging, especially on intergeneric and intraspecific delimitation. In that respect, the polyphasic approach has been widely used in taxonomic studies of cyanobacteria as an efficient method to identify new taxa and reassessing of those already established [4, 14]. This approach mainly integrates molecular data (*e.g.* 16S rRNA), morphology, ultrastructure, biochemistry and ecological aspects. The recent inclusion of new molecular biology tools has been particularly useful, as the secondary structure of the internal transcribed spacer (ITS) located between the 16S and 23S rRNA genes [15–18].

The integration of these tools has helped in the resolution of countless cryptogenera, which present distinct phylogenetically well-defined clades, but practically impossible to be separated only by morphology [4], clearly showing that genotypic diversity exceeds phenotypic diversity. Thereby, the polyphasic approach allows greater knowledge about polyphyletic genera [10, 19–26], contributing to a better systematic resolution of the world 'cyanoflora', a term that we propose herein to describe the diversity of Cyanobacteria into Phycoflora area.

However, the use of these tools is still not widely applied to most genera and species, especially those from tropical and subtropical environments, which has limited the real recognition of their biodiversity, as well as the knowledge about the cyanobacteria's evolutionary history and biogeography. This group has been neglected in the Brazilian marine floristic surveys, mainly for benthic environments, which are still poorly understood. Besides the difficulties in carrying out taxonomic studies due to the great problem in the definition of diacritic characteristics that assist in rapid and practical identification of taxa [27], it is important to highlight the small number of taxonomists working in this specific group in Brazil, which amplifies the significant knowledge gap about the cyanoflora.

2. Brazil: an extensive and environmentally diverse coastline

Brazil has 10.959 km of coastline (4°N to 33°S) which is bathed in its entire length by the Atlantic Ocean. Additionally, its coast has an Exclusive Economic Zone (EEZ) that includes up to 200 miles from the coast, encompassing the entire Continental Shelf and the oceanic islands [28]. The Brazilian coast is subdivided into three ecoregions based on the biogeographic system to classify the oceans: (1) Warm Temperate Southwestern Atlantic (South and Southeast regions); (2) Tropical Southwestern Atlantic (East and Northeast regions); and (3) North Brazil Shelf (Amazonia region) [29]. Included into the Tropical Southwestern Atlantic ecoregion are São Pedro & São Paulo Islands, Fernando de Noronha, Atoll das Rocas, Trindade and Martin Vaz Islands, and the Abrolhos Archipelago (**Figure 1**).

Horta et al. [30], based on the latitudinal temperature gradient occurring along the Brazilian coast, divided it into two large areas: (1) Tropical Region (covering the Northeast area); and (2) Warm Temperate Region (including the Southeast and South areas, with exception of Espírito Santo state, which is considered a transition zone) (**Figure 1**). The Tropical Region is characterized by the oligotrophic waters and its benthic phycoflora is found predominantly on sandy substrates. In the Warm Temperate Region, the benthic algae occur on rocky shores [30–32].

For macroalgae, the species richness presents a reduction in the north–south direction along the Brazilian coast [30]. The obtained data so far show this same pattern of species richness distribution for cyanobacteria. However, there are many

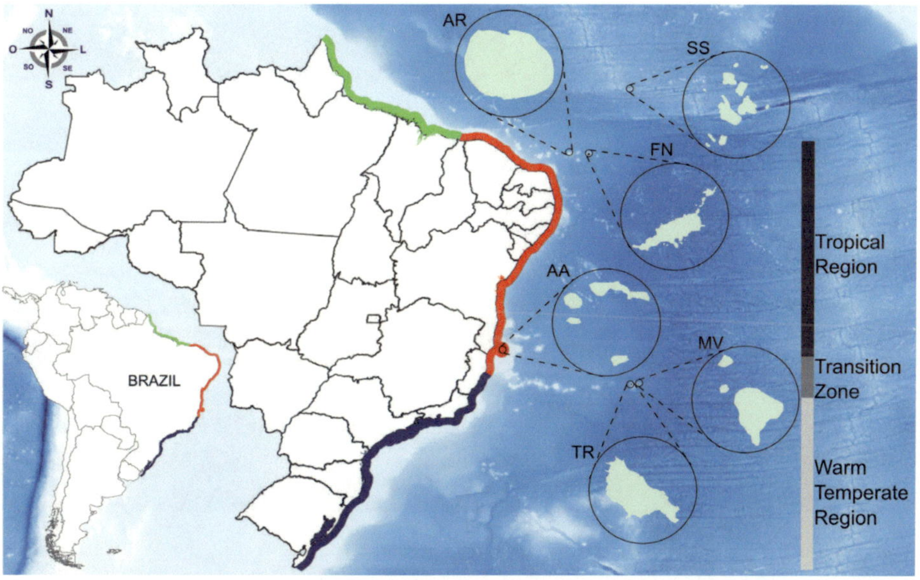

Figure 1.
Ecoregions (colored lines) based on Spalding et al. [29], and regions (gray scale bars) based on Horta et al. [30]. Green - North Brazil shelf; red - tropical southwestern Atlantic; blue - warm temperate southwestern Atlantic. AR – Atoll das Rocas; SS – São Pedro & São Paulo Islands; FN - Fernando de Noronha; TR - Trindade Island; MV - Martin Vaz Islands; AA - Abrolhos archipelago.

knowledge gaps about this group in the Brazilian coastline, which can modify this pattern observed up to now. Golubic et al. [33] and Hoffmann [34] highlighted that large areas with different climates can present a high cyanobacteria biodiversity, notably in the tropical zone, in which the great part of the Brazilian coast is found.

The Brazilian coastal region varies considerably in shape and width, including 3.000 km of coral reefs (0°50'S to 18°00'S) among which some are attached to the coast, and others are several kilometers offshore [35–37]. Charpy et al. [38] relate that benthic cyanobacteria constitute a major component of epiphytic, epilithic, and endolithic communities in reef ecosystems, performing an important role in these areas. Besides the coral reefs, the Brazilian coast presents several types of substrates, including beach rocky, sandstone formation, precambrian basement, and carbonate crusts, which can favor the occurrence of a great diversity of cyanobacteria (**Figure 2**).

In the marine benthic environments, the cyanobacteria can occupy the supralittoral, mediolitoral (intertidal zone) and infralittoral, which may grow in epilithic, epiphytic, and epizoic life forms [11, 39–41]. Cyanobacteria do not have specific morphological adaptations for fixation. Thus, these organisms typically occur associated with rough substrates, as beach rock which favors their adhesion, and microhabitats which present low hydrodynamics [41]. According to Taton and Hoffmann [42], these microorganisms can occur as endoliths in the locals with high water movement.

Regarding the vertical zonation in these environments, the morphophysiological characteristics of cyanobacteria play an important role in their distribution. Taton and Hoffmann [42] describe that these microorganisms can display zonation from supralittoral areas toward the intertidal zone, which is directly influenced by hydrodynamics, duration and frequency of subaerial exposure, and the type and amount

Figure 2.
General aspect of beaches on the Brazilian coast. **A.** *Praia do Francês (TSA);* **B.** *Paripueira (TSA);* **C.** *Pirambúzios (TSA);* **D.** *Pipa (TSA);* **E.** *Porto de Galinhas (TSA);* **F.** *Gaibú (TSA);* **G.** *Caponga (TSA);* **H.** *Itaqui (TSA);* **I.** *Araçagi (NBS);* **J.** *Pontal de Guaibura (WTSA);* **K.** *Praia das Focas (WTSA);* **L.** *Praia de Tassimirim (TSA);* **M.** *Ponta do Mutá (TSA);* **N.** *Praia do Francês (TSA);* **O.** *Imbassaí (TSA);* **P.** *Praia dos Castelhanos (WTSA);* **Q.** *Ponta das Canas (WTSA);* **R.** *Ponta da Nhá Pina (WTSA);* **S.** *Praia Grande (WTSA);* **T.** *Búzios (WTSA). (WTSA) Warm Temperate Southwestern Atlantic ecoregion (south and southeast regions); (TSA) Tropical Southwestern Atlantic ecoregion (east and northeast regions); (NBS) North Brazil Shelf ecoregion (subregion Amazonia region).*

of sediments. In that respect, the presence of heterocytes can favor the survival of these microorganisms in microhabitats with limited nutrient availability, as the supralittoral. Tomitani et al. [43] and Sohm et al. [44] related the predominance of Nostocales taxa in this abovementioned microhabitat, which has limiting abiotic variables.

On the other hand, homocyted taxa present high success in several microhabitats in the mediolitoral region, which remain in contact with water for a period of the day [41]. Besides the aforementioned zones, the infralittoral region is also understudied in the Brazilian marine environments, with only one species described for this zone, *Symploca infralitoralis*, Caires et al. [40], highlighting the need for greater sampling and analysis efforts in this environment.

3. How much is Brazilian marine benthic cyanoflora known?

In Brazil, even with the great development of phycological studies, the knowledge about the phylum Cyanobacteria has not evolved to the same degree, especially regarding the marine benthic environments [41, 45]. According to Komárek [46], only 5 to 10% of the diversity of cyanobacteria is known in the world. Menezes et al. [47] carried out a data compilation and registered the following species richness for the Brazilian marine environments: North - one taxon; Northeast - 45 taxa; Southeast - 91 taxa; and South - 16 taxa. However, these numbers include the diversity of planktonic cyanobacteria, which are more widely known than benthic ones.

Regarding the benthic taxa for the Brazilian coastal region, only 46 species are reported by 'Flora do Brasil 2020 Database', which are distributed as follows: 25 species for Oscillatoriales (homocyted taxa); six species for Nostocales (heterocystous taxa); two for Spirulinales (homocyted taxa); seven for Synechococcales (five homocyted taxa; and two single-celled taxa); four species for Chroococcales (single-celled taxa); and two species for Pleurocapsales (single-celled taxa) [48].

Cyanobacteria is a group of great ecological, health, economic and biotechnological interest nevertheless studies approaching benthic marine cyanobacteria in Brazil are scarce (**Figure 3**). Among the realized studies, most of them were developed in the southeastern region, mainly in the coastal environments of the São Paulo and Rio de Janeiro states, both included in Warm Temperate Region [49–65].

For the other Brazil's regions, the following studies are reported: South – Coutinho et al. [66] for Santa Catarina state, and Garcia-Baptista & Baptista [67] for Rio Grande do Sul state; and Northeast – Nogueira and Ferreira-Correia [68] for Maranhão, Branco et al. [69] for Pernambuco, and Caires et al. [10, 40, 41] for Bahia. Additionally, Caires et al. [70] described one new genus, *Neolyngbya* T.A. Caires, C.L. Sant'Anna et J.M.C. Nunes, which includes six species widespread for the Brazilian coast. Among them, *N. granulosa* T.A. Caires, C.L. Sant'Anna et J.M.C. Nunes was recently reevaluated based on genetic data by Lefler et al. [26], who proposed a new combination *Affixifilum granulosum* (Caires, Sant'Anna et Nunes) Lefler, D.E.Berthold et Laughinghouse.

Besides *Neolyngbya*, other two new genera were described for Brazilian coast in the last decade: *Capilliphycus* T.A. Caires, C.L. Sant'Anna et J.M.C. Nunes [10], including two species (*C. salinus* T.A. Caires, C.L. Sant'Anna et J.M.C. Nunes, and *C. tropicalis* T.A. Caires, C.L. Sant'Anna et J.M.C. Nunes); and *Halotia* D.B. Genuário *et al.*, with one species for marine benthic environment (*H. branconii* D.B. Genuário *et al.*). The distribution of the ten new species, including *Symploca infralitoralis* described for the Brazilian infralittoral, is showed in **Figure 4**.

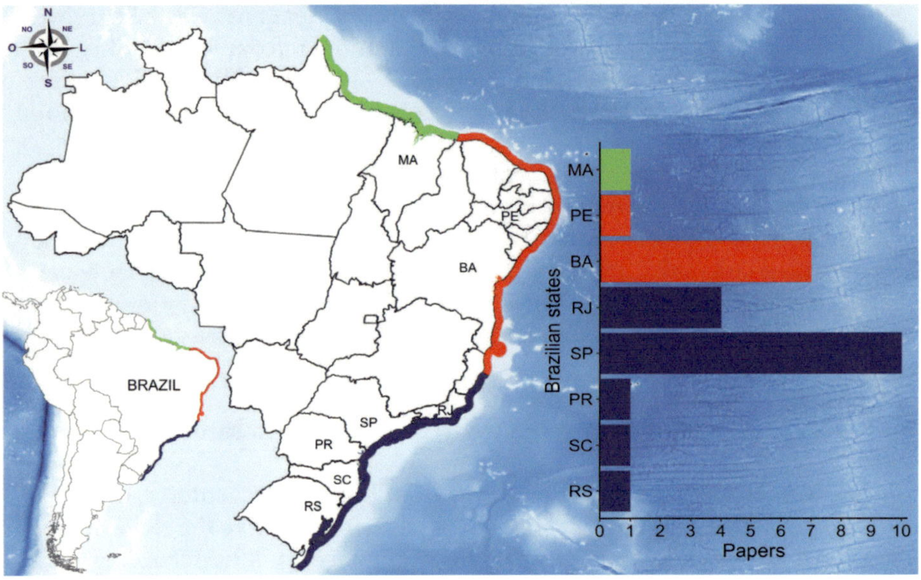

Figure 3.
Number of papers about benthic cyanobacteria developed in each coastal region Brazilian coastline: Green - North Brazil shelf (one study); red - tropical southwestern Atlantic (eight studies); blue - warm temperate southwestern Atlantic (seventeen studies). Brazilian states: MA – Maranhão; PE – Pernambuco; BA – Bahia; RJ – Rio de Janeiro; SP – São Paulo; PR - Paraná; SC – Santa Catarina; RS – Rio Grande do Sul.

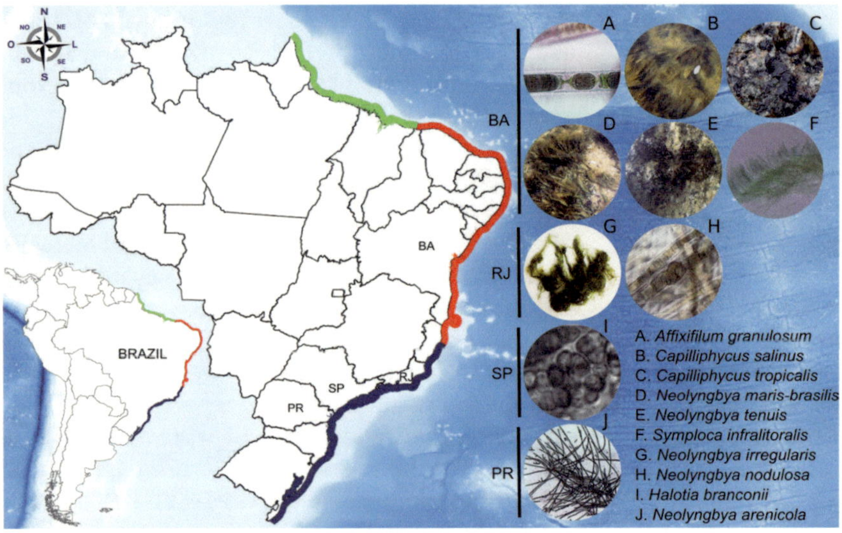

Figure 4.
Spatial distribution along the Brazilian coast of the new described genera and species in the last decade. Data based on Caires et al. [10, 40, 70] and Genuário et al. [64]. Green - North Brazil shelf (no taxon); red - tropical southwestern Atlantic (six taxa); blue - warm temperate southwestern Atlantic (four taxa). Brazilian states: BA – Bahia; RJ – Rio de Janeiro; SP – São Paulo; PR - Paraná.

Since the first published research about Brazilian marine benthic cyanobacteria which was carried out almost four decades ago, a limited number of studies was realized in this country (**Figure 5**). The number of studies and the interval among them demonstrate that the knowledge about the diversity and distribution of cyanobacteria are still incipient, especially when considering the coastal extension of Brazil and the great diversity of favorable habitats for the development of these organisms [41].

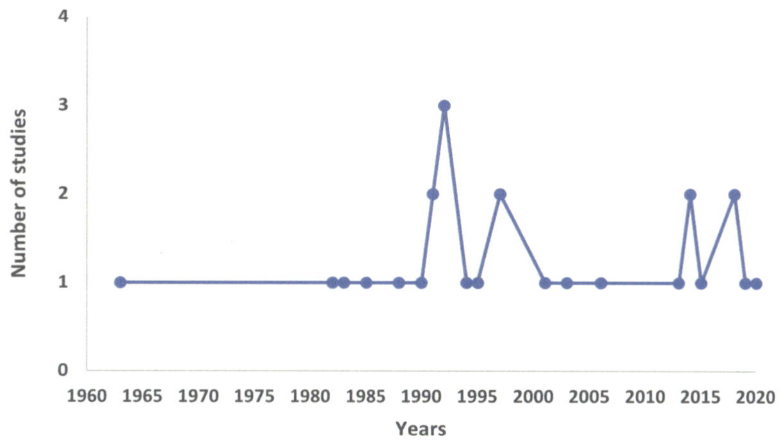

Figure 5.
Distribution of all studies carried out about marine benthic cyanobacteria in the Brazilian environments over time.

Furthermore, Brazil has some islands, as São Pedro & São Paulo, Fernando de Noronha, Atoll das Rocas, and Trindade and Martin Vaz, which are completely unknown about their cyanoflora, underestimating the biodiversity of this group. For Abrolhos Archipelago, only the studies carried out by Walter et al. [71] and Walter et al. [72] describe a new genus, *Adonisia* Walter et al., and a new species *Adonisia turfae* Walter et al., respectively. However, the descriptions of new taxa are not valid according to the rules of the International Code of Nomenclature of Prokaryotes - ICNP [73] and the International Code of Nomenclature for Algae, Fungi and Plants - Shenzhen Code [74]. Therefore, these taxa were not included in our analyses about benthic cyanobacteria diversity.

Regarding the relevance of ecological aspects related to benthic cyanobacteria, it is important to highlight that these microorganisms also are responsible for the "Black Band Disease" in the reef-forming corals, causing a direct impact on the functioning of these systems, and affecting their entire associated biota [75]. Some filamentous species have formed recurrent blooms, as observed in Abrolhos Archipelago by Ribeiro et al. [76]. According to Taylor et al. [77], the frequency of these blooms tends to increase globally as marine ecosystems undergo a process of eutrophication and thermal anomalies, which can be associated with global climate changes. These factors make evident the importance of studies that contribute to the knowledge of cyanobacteria, as well as about the interactions between this group and other organisms that occurred in the Brazilian marine habitats, supporting initiatives for the conservation of the reef environments.

4. Underestimated biotechnological potential of the Brazilian cyanoflora

Cyanobacteria presents a great physiological ability to survive under several abiotic conditions. This capacity is possibly related to a large number of secondary metabolites biosynthesized by different biochemical routes in these individuals [78]. These compounds are distributed into 260 cyanobacterial metabolite families and in ten different chemical classes, like lipopeptides, peptides, lipids, terpenes, polysaccharides, alkaloids, polyketides, macrolides/lactones, and indole compounds, [79, 80]. Among marine cyanobacteria, the filamentous forms usually have a greater quantity of natural bioactive products, with approximately 800 compounds [81].

Therefore, these microorganisms are evaluated as a rich source of biologically active secondary metabolites with potential biotechnological application [82, 83], especially in the pharmacological area, presenting substances with antitumor, antibacterial, antiviral, antifungal, antioxidant, anti-inflammatory and anticholinesterase activities [84–102].

According to Demay et al. [80], there are more than 90 genera of cyanobacteria that produce compounds with potential beneficial activities, most of them belonging to the orders Oscillatoriales, Nostocales, Chroococcales, and Synechococcales. However, the orders Pleurocapsales, Chroococcidiopsales, and Gloeobacterales remain poorly explored about their bioactivity potential. In that respect, the marine species, specifically the complex *Lyngbya-Moorena* genera stand out for the large production of bioactive metabolites.

Although Brazil has a long coastline, showing itself as a potential source of natural products, the biotechnological properties of Brazilian marine cyanobacteria are still almost completely unknown, underestimating one of the most diverse groups with promising potential for the possibility of isolating new biochemically active compounds. In Brazil, only the studies conducted by Caires et al. [103], Vaz [62], Silva et al. [65], and Armstrong et al. [104] deal with this theme, which demonstrated the great biotechnological potential of the Brazilian marine benthic cyanobacteria, highlighting the capacity of this group.

Approaches including biofertilization and nutraceutical applications have not yet been carried out with these Brazilian marine microorganisms. For biofuel production, only the study realized by Da Rós et al. [105] with a unicellular marine strain *Chlorogloea* sp. is registered. This research evaluated the chemical and physico-chemical properties of lipids obtained from distinct cyanobacterial strains for biodiesel production.

Therefore, the extensive Brazilian coastline is almost entirely unidentified regarding the biotechnological potential of the marine cyanobacteria occurring in this environment, underestimating the biochemiodiversity of one of the most promising groups concerning the possibility of isolating new biochemically active compounds with unprecedented molecular skeletons. Thus, the use of natural compounds, as those obtained from marine cyanobacteria, has become a viable alternative for the discovery of substances for the treatment of infections caused by resistant microorganisms [106], as well as the cyanobacterial biomass can become an important source to nutraceutical, biofertilization, and biofuel applications.

5. Future perspectives about Brazilian cyanoflora

The knowledge about the biodiversity of marine benthic cyanobacteria from the Brazilian coast is notoriously underestimated. The recognition of their cyanoflora makes an important contribution to the understanding of the functioning of marine ecosystems, as well as the ecological relationships in which cyanobacteria are present, such as coral reefs. Besides, the generation of scientific knowledge about marine biodiversity is one of the goals of the United Nations Decade for Ocean Science for Sustainable Development, providing data for monitoring and conservation of these environments, in addition to the construction of future public policies [107].

The number of new taxa related to the Brazilian coast in the last decade emphasizes the challenge of conducting further floristic surveys in the underexplored marine environments in order to fill an important lacune in the cyanoflora knowledge, as well as their biogeographic distribution. The future data set based on a polyphasic approach about this group can contribute to the definition of morphological markers for better delimitation of marine species, providing the basis for understanding evolutionary relationships and sustaining systematic decisions into Cyanobacteria. Furthermore, the recognition of this diversity supports studies that can reveal new bioactive compounds through the biotechnological prospection, aggregating value to these microorganisms.

Author details

Taiara A. Caires[1]* and Helen Michelle de J. Affe[2,3]

1 Programa de Pós-Graduação em Botânica, Departamento de Ciências Biológicas, Universidade Estadual de Feira de Santana, Feira de Santana, Bahia, Brazil

2 Faculdade de Oceanografia – FAOC, Universidade do Estado do Rio de Janeiro, Rua São Francisco Xavier, Maracanã, Rio de Janeiro, Brazil

3 Sub-Rede Oceanos, Rede Clima, Instituto Nacional de Pesquisas Espaciais - INPE, São José dos Campos, São Paulo, Brazil

*Address all correspondence to: taiaracaires@gmail.com

References

[1] Wilmotte A. Molecular Evolution and Taxonomy of the Cyanobacteria. *In*: Bryant, D. A. (ed). The Molecular Biology of Cyanobacteria. Kluwer Academi Pub. 1994; 1-25

[2] Golubic S, Seong-Joo L. Early Cyanobacterial fossil record: preservation palaeoenvironments and identification. European Journal Phycology 1999; 34:339-348

[3] Jill AS, Webb EA, Capone DG. Emerging patterns of marine nitrogen fixation. Nature Reviews Microbiology. 2011; 9:499-508

[4] Komárek J. A polyphasic approach for the taxonomy of cyanobacteria: principles and applications. European Journal of Phycology. 2016. DOI: 10.1080/09670262.2016.1163738

[5] Schopf JW, Packer BM. Early Archean (3.3-Billion to 3.5-Billion-Year-Old) Microfossils from Warrawoona Group, Australia. Science. 1987; 3:237:70-3

[6] Brocks JJ, Logan GA, Buick R, Summons RE. Archean Molecular Fossils and the Early Rise of Eukaryotes. Science. 1999; 285:1033-1036

[7] Komárek J. Cyanobacterial taxonomy: current problems and prospects for the integration of traditional and molecular approaches. Algae. 2006; 21:349-375

[8] Komárek J, Anagnostidis K. Cyanoprokaryota-2. Teil/2nd part: Oscillatoriales, in: Büdel, B., Krienitz, L., Gärtner, G., Schagerl, M. (Eds.), Süsswasserflora von Mitteleuropa 19/778 2. Elsevier/Spektrum, Heidelberg. 2005. 759p.

[9] Chen M, Schliep M, Willows RD, Cai Z, Neilan BA, Scheer H. A Red-Shifted Chlorophyll. Science. 2010; 329:1318-1319

[10] Caires TA, Lyra GM, Hentschke GS, Pedrini AG, Sant'Anna CL, Nunes JMC. *Neolyngbya* gen. nov. (Cyanobacteria, Oscillatoriaceae): A new filamentous benthic marine taxon widely distributed along the Brazilian coast. Molecular Phylogenetics and Evolution. 2018a; 120:196-211. ISNN 10557903

[11] Sant'Anna CL. Cyanophyceae marinhas bentônicas do Parque Estadual da Ilha do Cardoso, São Paulo, Brasil. Hoehnea. 1995; 22 (1/2):197-216

[12] Dvořák P, Jahodářová E, Hašler P, Gusev E, Pulíčková A. A new tropical cyanobacterium *Pinocchia polymorpha* gen. et sp. nov. derived from genus Pseudanabaena. *Fottea*. 2015; 15:113-120.

[13] Sant'Anna CL, Azevedo MTP, Agujaro LF. Manual Ilustrado e Contagem de Cianobactérias Planctônicas de Águas Continentais Brasileiras. 1ª ed. Rio de Janeiro. Interciência. 2006; 58p.

[14] Hoffmann L, Komárek J, Kaštovský J. System of cyanoprokaryotes (Cyanobacteria) – state in 2004. Algological Studies. 2005; 117: 95-115.

[15] Iteman I, Rippka R, De Marsac NT, Herdman M. rDNA analyses of planktonic heterocystous cyanobacteria, including members of the genera *Anabaenopsis* and *Cyanospira*. Microbiology. 2002; 148:481-496.

[16] Boyer SL, Flechtner VR, Johansen JR. Is the 16S–23S rRNA internal transcribed spacer (ITS) region a good tool for use in molecular systematics and population genetics? A case study in cyanobacteria. Mol. Biol. Evol. 2001; 18:1057-69.

[17] Boyer SL, Johansen JR, Flechtner VR. Phylogeny and genetic variance in terrestrial *Microcoleus*

(Cyanophyceae) species based on sequence analysis of the 16s rRNA gene and associated 16S–23S ITS region. J. Phycol. 2002; 38:1222-1235.

[18] Johansen JR, Kovacik L, Casamatta DA, Fučiková K, Kaštovský J. Utility of 16S–23S ITSsequence and secondary structure for recognition of intrageneric and intergeneric limits within cyanobacterial taxa: *Leptolyngbya corticola* sp. nov. (Pseudanabaenaceae, Cyanobacteria). Nova Hedwigia. 2011; 92:283-302.

[19] Sharp K, Arthur KE, Gu L, Ross C, Harrison G, Gunasekera SP, Meickle T, Matthew S, Luesch H. *et al.* Phylogenetic and chemical diversity of three chemotypes of bloom-forming *Lyngbya* species (Cyanobacteria: Oscillatoriales) from reefs of southeastern Florida. Appl. Environ. Microbiol. 2009; 75:2879-2888.

[20] Engene N, Rottacker EC, Kastovsky J, Byrum T, Choi H, Ellisman MH, Komárek J, Gerwick WH. *Moorea producta* gen. nov., sp. nov. and *Moorea bouillonii* comb. nov., tropical marine cyanobacteria rich in bioactive secondary metabolites. Int. Journ. of Systematic and Evolutionary Microbiology. 2012; 62:202-214.

[21] Engene N, Gunasekera SP, Gerwick WH, Paul VJ. Phylogenetic Inferences Reveal a Large Extent of Novel Biodiversity in Chemically Rich Tropical Marine Cyanobacteria. Applied and Environmental Microbiology. 2013; 79 (6):1882-1888.

[22] Engene N, Tronholm A, Paul VJ. Uncovering cryptic diversity of Lyngbya: the new tropical marine cyanobacterial genus *Dapis* (Oscillatoriales). J Phycol. 2018; 54(4):435-446.

[23] Komárek J, Zapomělová E, Šmarda J, Kopecký J, Rejmánková E, Woodhouse J, Neilan BA, Komárková J. Polyphasic evaluation of *Limnoraphis robusta*, a water-bloom forming cyanobacterium from Lake Atitlán, Guatemala, with a description of *Limnoraphis* gen. nov. Fottea. 2013; 13(1):39-52.

[24] Mcgregor GB, Sendall BC. Phylogeny and toxicology of *Lyngbya wollei* (Cyanobacteria, Oscillatoriales) from north-eastern Australia, with a description of *Microseira* gen. nov. J. Phycol. 2014; 51(1):109-119.

[25] Stoyanov P, Moten D, Mladenov R, Dzhambazov B, Teneva I. Phylogenetic Relationships of Some Filamentous Cyanoprokaryotic Species. Evolutionary Bioinformatics. 2014; 10: 39-49.

[26] Lefler F, Berthold DE, Laughinghouse IV HD. The occurrence of *Affixifilum* gen. nov. and *Neolyngbya* (Oscillatoriaceae) in South Florida (USA), with the description of *A. floridanum* sp. nov. and *N. biscaynensis* sp. nov. Journal of Phycology. 2020. DOI: 10.1111/jpy.13065

[27] Kling HJ, Watson SA. New planktic species of *Pseudanabaena* (Cyanoprokaryota, Oscillatoriales) from North American large lakes. Hydrobiologia. 2003; 502: 383-388

[28] Castro BM, Brandini FP, Fortes JF. A Amazônia Azul: recursos e preservação. Revista USP. 2017; 113:7-26. https://doi.org/10.11606/issn.2316-9036.v0i113p7-26

[29] Spalding MD, Fox HE, Allen GR, Davidson N, Ferdaña ZA, Finlayson M, Halpern BS, Jorge MA, Lombana A, Lourie AS, Martin KD, Mcmanus E, Molnar J, Recchia CA, Robertson J. Marine Ecoregions of the World: A Bioregionalization of Coastal and Shelf Areas. 2007; 57(7):573-583

[30] Horta PA, Amancio E, Coimbra CS, Oliveira EC. Considerações sobre a distribuição e origem da flora de

macroalgas marinhas brasileiras. Hoehnea. 2001; 28(3):243-265

[31] Oliveira Filho EC. Algas marinhas bentônicas do Brasil. Tese de Livre-Docência, Universidade de São Paulo, São Paulo. 1977; 407p.

[32] Castro BM, Miranda LB. Physical oceanography of the western atlantic continental shelf located between 4°N and 34°S: Coastal segment (4,W). *In*: The Sea. 1998; 11:1

[33] Golubic S, Le Campion-Alsumard T, Campbell SE. Diversity of Marine Cyanobacteria. *In*: Marine Cyanobacteria. Charpy, L., Larkum, A., W., D., Eds. Monaco Musee Océanographique, Monaco, 1999; 19, pp. 53-76

[34] Hoffmann L. Marine cyanobacteria in tropical regions: diversity and ecology. Eur J Phycol. 1999; 34:371-379

[35] Leão ZMAN, Kikuchi RKP, Testa V. Corals and coral reefs of Brazil. *In* Cortes J. (ed.). Latin American Coral Reefs. Amsterdam: Elsevier Publisher. 2003. p. 9-52

[36] Ferreira BP, Costa MBSF, Coxey MS, Gaspar ALB, Veleda D, Araujo M. The effects of sea surface temperatures anomalies on oceanic coral reef systems in the southwestern tropical Atlantic. Coral Reefs. 2013; 32(2): 441-454

[37] Leão ZMAN, Kikuchi RKP, Ferreira BP, Neves EG, Sovierzoski HH, Oliveira MDM, Maida M, Correia MD, Johnsson R. Brazilian coral reefs in a period of global change: A synthesis. Brazilian Journal of Oceanography. 2016; 64(spe2), 97-116

[38] Charpy L, Casaerto BE, Langlade MJ, Suzuki Y. Cyanobacteria in coral reef ecosystems: a review. J Mar Biol. 2012; 2012:1-9

[39] Graham LE, Wilcox LW. Algae. Prentice Hall, New Jersey. 2000; pp. 97-131

[40] Caires TA, Sant'Anna CL, Nunes JMC. A new species of marine benthic cyanobacteria from the infralittoral of Brazil: *Symploca infralitoralis* sp. nov. Brazilian Journal of Botany. 2013; 36:159-163

[41] Caires TA, Sant'Anna CL, Nunes JMC. Biodiversity of benthic filamentous Cyanobacteria in tropical marine environments of Bahia State, Northeastern Brazil. Brazilian Journal of Botany. 2019; 42:149-170

[42] Taton A, Hofmann L. Marine cyanophyceae of Papua New Guinea. VII Endoliths. Algol. Stud. 2003; 109: 537-554.

[43] Tomitani A, Knoll AH, Cavanaugh CM, Ohno T. The evolutionary diversification of cyanobacteria: molecular–phylogenetic and paleontological perspectives. PNAS. 2006; 103:5442-5447

[44] Sohm JA, Webb EA, Capone DG. Emerging patterns of marine nitrogen fixation. Microbiol. 2011; 9:499-508

[45] Crispino LMB. Cianobactérias Marinhas Bentônicas do Estado de São Paulo. Tese de Doutorado em Biodiversidade Vegetal e Meio Ambiente - Instituto de Botânica de São Paulo. São Paulo. 2007; 154p.

[46] Komárek J. Planktic oscillatorialean cyanoprokaryotes (short review according to combined phenotype and molecular aspects). Hydrobiologia. 2003; 502:367-382

[47] Menezes M. *et al.* Update of the Brazilian floristic list of Algae and Cyanobacteria. Rodriguésia. 2015; 66 (4): 1047-1062

[48] Flora do Brasil 2020 Database - available at: http://floradobrasil.jbrj.gov.br/reflora/listaBrasil

[49] Tórgo FMS. Observações sobre *Ralfsia expansa* (Phaeophyceae) e *Hyella*

caespitosa (Cyanophyceae). Boletim do Museu Nacional, Nova Série, Botânica, Rio de Janeiro. 1963; n° 30.

[50] Baeta Neves MHC. Étude dês Cyanophycées marines de la région de Cabo Frio (Rio de Janeiro – Brésil): taxonomie et essai d'interpretation écologique. Tese de Doutorado, Université Pierre et Marie Curie, Paris. 1988; 155p.

[51] Baeta Neves MHC. Estudo das Cianofíceas Marinhas Bentônicas da Região de Cabo Frio (Estado do Rio de Janeiro, Brasil). I – Chroococcales. Hoehnea. 1991; 18(1): 191-204

[52] Baeta Neves MHC. Estudo das Cianofíceas Marinhas Bentônicas da Região de Cabo Frio (Estado do Rio de Janeiro, Brasil). II – Hormogonae. Revista Brasileira de Biologia. 1992; 52(4): 641-659.

[53] Baeta Neves MHC, Casarin AJ. As cianofíceas das salinas de Cabo Frio – Brasil. Acta Biologica Leopoldensia. 1990; 12(1):99-123

[54] Baeta Neves MHC. Tribuzi D. Les Cyanophycées de la mangrove de la "Ponta do Pai Vitório" de la região de Cabo Frio (RJ, Brésil). Acta Biológica Leopoldensia. 1992; 14:29-52

[55] Sant'Anna CL, Simonetti C. Cianobactérias marinhas bentônicas das Praias de Peruíbe edos Sonhos, Município de Itanhaém, São Paulo, II: espécies epilíticas e epizoicas. Revista Brasileira de Biologia. 1991; 52(3):515-523

[56] Sant'Anna CL. Cyanophyceae marinhas bentônicas da região de Ubatuba, SP, Brasil. Hoehnea. 1997; 24: 57-74

[57] Branco LHZ, Sant'Anna CL, Azevedo MTP, Sormus L. Cyanophyte flora from Cardoso Island, São Paulo State, Brazil, 2: Oscillatoriales. Algological Studies. 1997; 84:39-52

[58] Crispino LMB, Sant'Anna CL. Cianobactérias marinhas bentônicas de ilhas costeiras do Estado de São Paulo, Brasil. Revista Brasil. Bot. 2006; 29(4):639-655

[59] Sant'Anna CL, Bicudo RMT, Pereira HASL. Nostocophyceae (Cyanophyceae) do Parque Estadual da Ilha do Cardoso, Estado de São Paulo, Brasil. Rickia, 1983; 10: 1-27

[60] Sant'Anna CL, Cordeiro-Marino M, Braga MRA, Guimarães SMPB. Cianofíceas marinhas bentônicas das Praias de Peruíbe e dos Sonhos, Município de Itanhaém, São Paulo, Brasil. Rickia. 1985; 12:89-112

[61] Sant'Anna CL, Azevedo MTP, Branco LHZ, Braga MRA, Cordeiro-Marino M, Guimarães SMPB. Cianofíceas marinhas bentônicas das praias de Peruíbe e dos Sonhos, Município de Itanhaém, SP, Brasil, III. Revista Brasileira de Biologia. 1994; 55:389-407

[62] Vaz MGMV, GenuárioDB,AndreoteAPD,MaloneCFS, Sant'Anna CL, Barbiero L, Fiore MF. *Pantanalinema* gen. nov. and *Alkalinema* gen. nov.: novel pseudanabaenacean genera (Cyanobacteria) isolated from saline–alkaline lakes. Int. J. Syst. Evol. Microbiol. 2014; 65: 298-308

[63] Silva CSP, Genuário DB, Vaz MGMV, Fiore MF. Phylogeny of culturable cyanobacteria from Brazilian mangroves. Syst Appl Microbiol, 2014; 37: 100-112

[64] Genuário DB, Vaz MGMV, Hentschke GS, Sant'Anna CL, Fiore MF. *Halotia* gen. nov., a phylogenetically and physiologically coherent cyanobacterial genus isolated from marine coastal environments. Int J Syst Evol Microbiol. 2015; 65:663-675

[65] Silva EM, Vaz MGMV, Genuário DB, Armstrong L, Fiore MF, Debonsi HM.

Novel marine cyanobacteria from the Atlantic coast of Brazil, Applied Phycology. 2020; 1:1, 58-72.DOI: 10.1080/26388081.2020.1753571

[66] Coutinho R. Taxonomia, distribuição, crescimento sazonal, reprodução e biomassa das algas bentônicas do estuário da Lagoa dos Patos (RS). Dissertação de Mestrado, Universidade do Rio Grande, Rio Grande. 1982; 232p.

[67] Garcia-Baptista M, Baptista LRM. Algas psâmicas de Jardim Beira Mar, Capão da Canoa, Rio Grande do Sul. Revista Brasileira de Biologia. 1991; 52(2):325-342

[68] Nogueira NMC, Ferreira-Correia MM. Cyanophyceae/ Cyanobacteria in red mangrove Forest at Mosquitos and Coqueiros estuaries, São Luiz, State of Maranhão, Brazil. Brazilian Journal of Biology. 2001; 61: 347-356

[69] Branco LHZ, Moura NA, Silva AC, Bittencourt-Oliveira MC. Biodiversidade e considerações biogeográficas das Cyanobacteria de uma área de manguezal do estado de Pernambuco, Brasil. Acta Bot. Bras. 2003; 17: 585-596

[70] Caires TA, Lyra GM, Hentschke GS, Silva AMS, Araújo VL, Sant' Anna CL, Nunes JMC. Polyphasic delimitation of a filamentous marine genus, *Capillus* gen. Nov. (Cyanobacteria, Oscillatoriaceae) with the description of two Brazilian species. Algae. 2018b; 33(4):291-304

[71] Walter JM, Coutinho FH, Dutilh BE, Swings J, Thompson FL, Thompson CC. Ecogenomics and Taxonomy of Cyanobacteria Phylum. Front. Microbiol. 2017; 8:2132. DOI: 10.3389/fmicb.2017.02132

[72] Walter JM, Coutinho FH, Leomil L, Hargreaves PI, Campeão ME, Vieira VV, Silva BS, Fistaro GO, Salomon PS, Sawabe T, Mino S, Hosokawa M, Miyashita H, Maruyama F, van Verk MC, Dutilh BE, Thompson CC, Thompson FL. Ecogenomics of the Marine Benthic Filamentous Cyanobacterium *Adonisia* Microbial Ecology. 2019. https://doi.org/10.1007/s00248-019-01480-x

[73] Parker CT, Tindall BJ, Garrity GM (Editors). International Code of Nomenclature of Prokaryotes Prokaryotic Code (2008 Revision) 2019, volume 69, issue 1A, pages S1–S111. Int J Syst Evol Microbiol 2019; 69:S1

[74] Turland, NJ, Wiersema JH, Barrie FR, Greuter W, Hawksworth DL, Herendeen PS, Knapp S, Kusber WH, Li DZ, Marhold K, May TW, McNeill J, Monro AM, Prado J, Price MJ, Smith GF. International Code of Nomenclature for algae, fungi, and plants (Shenzhen Code) Regnum Vegetabile 159. Glashütten: Koeltz Botanical Books. 2018.

[75] Sato Y, Ling ES, Turaev D, Laffy P, Weynberg K, Rattei T, Willis B, Bourne D. Unraveling the microbial processes of black band disease in corals through integrated genomics. Scientific Reports. 2017; 7:1-14

[76] Ribeiro F. V., Padula, V., Moura, R. L., Moraes, F. C., Salomon, P. S., Gibran, F. Z., Motta, F. S., Pereira, R. C., Amado-Filho, G. M. & Pereira-Filho, G. H. 2017. Massive opisthobranch aggregation in the largest coralline reefs in the South Atlantic Ocean: are mesoherbivores underestimated top-down players? Bulletin of Marine Sciences 93 (3): 915-916.

[77] Taylor M, Stahl-Timmins W, Redshaw C, Osborne N. Toxic alkaloids in *Lyngbya majuscula* and related tropical marine cyanobacteria. Harmful Algae. 2014; 31:1-8

[78] Singh RK, Tiwari SP, Rai AK, Mohapatra TM. Cyanobacteria: an emerging source for drug discovery. The Journal of Antibiotics. 2011; 64:401-412

[79] Burja AM, Banaigs B, Abou-Mansour E, Burgess JG, Wright PC. Marine cyanobacteria – a profilic source of natural products. Tetrahedron. 2011; 57: 9347-9377

[80] Demay J, Bernard C, Reinhardt A, Benjamin M. Natural Products from Cyanobacteria: Focus on Beneficial Activities. Mar. Drugs. 2019; 17(6):320. https://doi.org/10.3390/md17060320

[81] Jones AC, Monroe EA, Eisman EB, Gerwick L, Sherman DH, Gerwick WH. The unique mechanistic transformations involved in the biosynthesis of modular natural products from marine cyanobacteria. Nat. Prod. Rep. 2010; 27:1048-1065

[82] Carmichael WW. The cyanotoxins - bioactive metabolites of cyanobacteria: occurrence, ecological role, taxonomic concerns and effects on humans. Journal of Phycology, 2001; 37(3):9-10

[83] Costa M, Costa-Rodrigues J, Fernandes ML, Barros P, Vasconcelos V, Martins R. Marine Cyanobacteria Compounds with Anticancer Properties: A Review on the Implication of Apoptosis. Marine Drugs. 2012; 10(10):2181-2207

[84] Mynderse JS, Moore RE, Kashiwagi M, Norton TR. Antileukemia Activity in the Osciliatoriaceae: Isolation of Debromoaplysiatoxin from *Lyngbya*. Science, 1977; 196: 538-540

[85] Harrigan GG, Goetzl G. Symbiotic and dietary marine microalgae as a source of bioactive molecules–experience from natural products research. Journal of Applied Phycology. 2002; 14(2):103-108

[86] Mian P, Heilmann J, Bürgi HR, Sticher O. Biological Screening of Terrestrial and Freshwater Cyanobacteria for Antimicrobial Activity, Brine Shrimp Lethality, and Cytotoxicity. Pharmaceutical Biology. 2003; 41(4): 243-247

[87] Surakka A, Sihvonen LM, Lehtimaki JM, Wahlsten M, Vuorela P, Sivonen K. Benthic Cyanobacteria from the Baltic Sea Contain Cytotoxic *Anabaena*, *Nodularia*, and *Nostoc* Strains and an Apoptosis-Inducing *Phormidium* Strain. Wiley Periodicals. 2005. DOI: 10.1002/tox.20119

[88] Gademann K, Portmann C. Secundary metabolites from Cyanobacteria: Complex Structures and Powerful Bioactivities. Current Organic Chemistry. 2008; 12: 326-341

[89] Martins RF, Ramos MF, Herfindal L, Sousa JA, Skærven K, Vasconcelos VM. Antimicrobial and Cytotoxic Assessment of Marine Cyanobacteria – *Synechocystis* and *Synechococcus*. Marine Drugs. 2008; 6(1):1-11

[90] Simmons TL, Niclas EN, Ureña LD. *et al.* Viridamides A and B, Lipodepsipeptides with Anti-Protozoal Activity from the Marine Cyanobacterium *Oscillatoria nigro-viridis*. J Nat Prod. 2008; 71(9):1544-1550

[91] Becher PG, Baumann HI, Gademann K, Jütner F. The cyanobacterial alkaloid nostocarboline: an inhibitor of acetylcholinesterase and trypsin. Journal of Applied Phycology. 2009; 21: 103-110

[92] Chlipala G, Shunyan M, Esperanza J, Carcache B, Ito A, Bazarek S, Orjala J. Investigation of antimicrobial and protease-inhibitory activity from cultured cyanobacteria. Pharmaceutical Biology. 2009; 47 (1):53-60

[93] Liu L, Rein KS. New Peptides Isolated from *Lyngbya* Species: A Review. Mar. Drugs. 2010; 8:1817-1837

[94] Nunnery JK, Mevers E, Gerwick WH. Biologically active secondary metabolites from marine

cyanobacteria. Curr Opin Biotechnol. 2010; 21:787-793

[95] Sanchez LM, Lopez D, Vesely BA *et al.* Almiramides A–C: Discovery and Development of a New Class of Leishmaniasis Lead Compounds. J Med Chem. 2010; 53(10):4187-4197

[96] Khalid MN, Shameel MM. Studies on the Bioactivity and Phycochemistry of *Lyngbya* Agardh (Cyanophycota). International Journal on Algae. 2011; 13 (3): 234-249

[97] Oftedal L, Skjærven KH, Coyne RT, Edvardsen B, Rohrlack T, Skulberg OM, Døskeland SO, Herfindal L. The apoptosis-inducing activity towards leukemia and lymphoma cells in a cyanobacterial culture collection is not associated with mouse bioassay toxicity. J Ind Microbiol Biotechnol. 2011; 38: 489-501

[98] Balunas MJ, Grosso MF, Villa FA, *et al.* Coibacins A-D, Anti-leishmanial Marine Cyanobacterial Polyketides with Intriguing Biosynthetic Origins. Org Lett. 2012; 14(15):3878-3881

[99] Hau AM, Greenwood JA, Lo CV, Serrill JD, Proteau PJ, Ganley IG, Mcphail KL, Ishmael JE. Coibamide A Induces mTOR-Independent Autophagy and Cell Death in Human Glioblastoma Cells. Plos One. 2013; 8(6):1-16

[100] Felczykowska A, Pawlik A, Mazur-Marzec H, Toruńska-Sitarz A, Narajczyk M, Richert M, Węgrzyn G, Herman-Antosiewicz A. Selective inhibition of cancer cells' proliferation by compounds included in extracts from Baltic Sea cyanobacteria. Toxicon. 2015; 108:1-10

[101] Soares AR, Engene N, Gunasekera SP, Sneed JM, Paul VJ. Carriebowlinol, an Antimicrobial Tetrahydroquinolinol from an Assemblage of Marine Cyanobacteria Containing a Novel Taxon. J. Nat. Prod. 2015; 78(3):534-538

[102] Gheda SF, Ismail GA. Natural products from some soil cyanobacterial extracts with potent antimicrobial, antioxidant and cytotoxic activities. An. Acad. Bras. Ciênc. 2020; 92(2):e20190934

[103] Caires TA, Silva AMS, Vasconcelos VM, Affe HMJ, Souza-Neta LC, Boness HVM, Sant'Anna CL, Nunes JMC. Biotechnological potential of *Neolyngbya* (Cyanobacteria), a new marine benthic filamentous genus from Brazil. Algal Research. 2018c; 36:1-9

[104] Armstrong L, Vaz MGMV, Genuário DB, Fiore MF, Debonsi HM. Volatile compounds produced by cyanobacteria isolated from mangrove environment. Current Microbiology. 2019; 76:575-582

[105] Da Rós PCM, Silva CSP, Silva-Estenico ME, Fiore MF, De Castro HF Assessment of Chemical and Physico-Chemical Properties of Cyanobacterial Lipids for Biodiesel Production. Marine Drugs. 2013; 11:2365-2381. https://doi.org/10.3390/md11072365

[106] Pandey VD. Cyanobacterial natural products as antimicrobial agents, Int. J. Curr. Microbiol. App. Sci. 2015; 4(1):310-317

[107] UNESCO. 2020. UNESCO - United Nations Educational, Scientific and Cultural Organization. United Nations Decade of Ocean Science for Sustainable Development (2021-2030) - available at: accessible at https://oceandecade.org/.